INFORMATION SYSTEMS AND COMPUTING TECHNOLOGY

PROCEEDINGS OF THE INTERNATIONAL CONFERENCE ON INFORMATION SYSTEMS AND COMPUTING TECHNOLOGY (ISCT 2013), WUXI, CHINA, 15–16 SEPTEMBER 2013

# Information Systems and Computing Technology

*Editors*

Lei Zhang

*Wuxi Innovation Center of Shanghai Institute of Technical Physics of the Chinese Academic of Sciences, Shanghai, China*

Yonggen Gu

*Huzhou Teachers College, School of Information and Engineering, Huzhou, China*

CRC Press
Taylor & Francis Group
Boca Raton London New York Leiden

CRC Press is an imprint of the
Taylor & Francis Group, an **informa** business
A BALKEMA BOOK

*CRC Press/Balkema is an imprint of the Taylor & Francis Group, an informa business*



Typeset by V Publishing Solutions Pvt Ltd., Chennai, India
Printed and bound in Great Britain by CPI Group (UK) Ltd, Croydon, CR0 4YY



Published by: CRC Press/Balkema
P.O. Box 11320, 2301 EH Leiden, The Netherlands
e-mail: Pub.NL@taylorandfrancis.com
www.crcpress.com – www.taylorandfrancis.com

ISBN: 978-1-138-00115-2 (Hbk)
ISBN: 978-1-315-85144-0 (eBook)

*Information Systems and Computing Technology – Zhang & Gu (eds)*
*© 2013 Taylor & Francis Group, London, ISBN 978-1-138-00115-2*

# Table of contents

*Information Systems and Computing Technology – Zhang & Gu (eds)*
*© 2013 Taylor & Francis Group, London, ISBN 978-1-138-00115-2*

# Preface

The information system is a complex system including data collecting, storing, processing and delivering. The main components of information systems are computer hardware and software, telecommunications, databases and data warehouses, human resources, and procedures. With the development of the information system, the innovation technologies and their applications continuously appear, such as the Internet of Things (IOT), cloud computing, big data and smart cities.

2013 International Conference on Information Systems and Computing Technology (ISCT 2013) will he held during 15–16 September 2013, in Wuxi, Jiangsu Province, China. ISCT 2013 aims to bring together researchers, scientists, engineers, and scholar students to exchange and share their experiences, new ideas, and research results, and discuss the practical challenges encountered and the solutions adopted.

We thank the Wuxi Innovation Center of Shanghai Institute of Technical Physics of the Chinese Academic of Sciences and School of Information and Engineering of Huzhou Teachers College. And also thank the Secretariat of the organizing committee and others for any help.

Last but not least, we would like to express our deep gratitude to all authors, reviewers for their excellent work, and Léon Bijnsdorp, Lukas Goosen, Richard Gundel and other editors from Taylor & Francis Group for their wonderful work.

*Information Systems and Computing Technology – Zhang & Gu (eds)*
*© 2013 Taylor & Francis Group, London, ISBN 978-1-138-00115-2*

# Sponsors

- Wuxi Innovation Center of Shanghai Institute of Technical Physics, CAS
- School of Information and Engineering of Huzhou Teachers College

*Invited papers*

*Information Systems and Computing Technology – Zhang & Gu (eds)*
*© 2013 Taylor & Francis Group, London, ISBN 978-1-138-00115-2*

# Incorporating the multi-cross-sectional temporal effect in Geographically Weighted Logit Regression

Kaizhao Wu
*Land and Resources Technology Center of Guangdong Province, Guangzhou, China*

Biao Liu & Bo Huang
*Department of Geography and Resource Management, The Chinese University of Hong Kong, Shatin, N.T., Hong Kong*

Z. Lei
*The Shanghai Institute of Technical Physics of the Chinese Academy of Sciences, Shanghai, China*

ABSTRACT: Sustainability concerns have aroused great interest among policy makers in the interrelated land use change systems. Land use change analysis becomes a hot topic. In recent years, considerable attention has been devoted to consider the spatial effect. Less attention has been paid to analyze the temporal effect. Learning how the temporal effect affect the accuracy of land use change model is thus of our great interesting. This paper incorporates the temporal effect in Geographically Weighted Logit Regression (GWLR), which can consider spatial nonstationarity in land use change modeling. Specifically, there will be two intervals to consider the temporal impacts in GWLR on land cover change modeling. Firstly, different time intervals in different land cover change process on one point will cause different results. Secondly, the other kind of time intervals is the time interval between the land use changes. In this study, for the first kind of temporal interval, we try to add an 'age' variable to the regression model. For the second, we proposed a new weighting function that combines the "geographical space" and "temporal space" between one observation and its neighbors, such that (1) neighbors with greater geographical distances from the subject point are assigned smaller weights, and (2) at a given geographical distance, neighboring points with less temporal interval to that of the subject point are assigned larger weights. The GWLR with weight function which consider both the two kind temporal effects will be implemented to be compared with the others. Case studies of land use change patterns in Calgary, Canada will be implemented. The results indicate that the GWLR model considering the temporal effect performs better than the other models.

## 1 INTRODUCTION

Land use change analysis, which becomes a hot topic recently, plays a key role in urban planning. In the past two decades, considerable effort has been made in regards to land use change modeling (Baker, 1989; Agarwal et al. 2002). Various land use change models (VeldKamp and Lambin, 2001; Berling-Wolff and Wu, 2004) have been developed to facilitate understanding of the complex interactions underlying the land use change.

Specifically, a variety of statistical methods such as Markov chain analysis (Lopez et al. 2001, Weng 2002), multiple regression analysis (Theobald and Hobbs 1998), are employed. Afterwards, Artificial Neural Networks (ANNs) (Pajanowski et al. 2002), Multi-agent system and Cellular Automata (CA) capable of dynamic spatial simulation and innovative bottom-up approach (Clarke and Gaydos 1998; Wu 1998; Wu 2002) are used. Despite different levels of success in their specific applications, their drawbacks limit their efficiency in land use change modeling. Markov chain analysis uses a transition matrix to describe the change of land use but cannot reveal the causal factors and their significance. In multiple regression analysis, the land use data often defied fundamental assumptions such as the normal distribution and model linearity (James and McCulloch 1990, Olden and Jackson 2001). Artificial neural network is a powerful method used to model nonlinear relationships but it is prone to overfitting the training data and cannot be relied upon to ensure the generalization performance (Sui, 1994). Moreover, ANNs have a static nature, in which causal factors are not dynamic. Multi-Agent System (MAS) is a microscopic simulation method and is therefore unable to fit the requirements of large scale modeling. Moreover, it is difficult to define the perception rule for the agent interactions in MAS (Parker et al. 2001). CA recognizes the temporal and spatial context of each cell; however, it can only share information across immediate cells, so that more dispersed interactions and correlations are largely ignored.

In order to identify socio-economic driven forces of land use and to predict corresponding land use, regression modeling plays a fundamental role in examining the relationship quantitatively between land use changes and explanatory variables. In such a framework, spatial non-stationarity, which means different relationships exist at different points in space, should be considered. It's a very important aspect on land use change analysis. Geographically Weighted Regression (GWR) for depicting the spatial nonstationarity in a regression has been developed. Meanwhile, McMillen and McDonald (1998) point out that one need apply standard logit or probit methods to the discrete data in place of least squares. Thus, Geographically Weighted Logit Regression (GWLR) is introduced into land cover/cover change modeling.

Now, more specific attentions begin to be paid on spatial-temporal analysis on land cover change (Huang et al. 2008). Temporal effect should also be incorporated into the GWRL model. Thus, this study aims to construct a novel statistical model in a GWLR framework to considering both spatial and temporal effects to assist the spatio-temporal land use change analysis. A case study on multi-temporal land use change in Calgary, Canada will be carried out to evaluate the performance and reliability of the proposed model.

## 2 METHODOLOGY

In order to consider the temporal effect, GWLR Model will be extended to Temporal GWLR (TGWLR) Model which considers both spatial and temporal nonstationarity. A significant improvement achieved by TGWLR will be shown by the comparison of estimation results of several models.

### 2.1 *Overview of GWLR model*

Suppose we have a set of n observations $\{x_{ij}\}$ with the spatial coordinates $\{(u_i,v_i)\}$, $i = 1, 2, \ldots, n$, on q predictor variables, $j = 1, 2, \ldots, q$, and a set of n observations on a dependent or response variable $\{y_i\}$. The underlying model for GWLR is:

$$y_i = \log_e\left(\frac{p}{1-p}\right) = a(u_i,v_i) + b_1(u_i,v_i)x_{i1} + b_2(u_i,v_i)x_{i2} + \cdots + b_q(u_i,v_i)x_{im} \tag{1}$$

where $\{a(u_i,v_i), b_1(u_i,v_i), b_2(u_i,v_i), \ldots, b_q(u_i,v_i)x_{im}\}$ are $q+1$ continuous functions of the location $(u_i,v_i)$ in the study area.

The parameter estimation is a moving window process. A region or window was drawn around a location i, and all the data points within this region or window were then used to estimate the parameters in eq. 1. The estimator of $b_i$ is given at each location i by using the ordinary logit model with *X*'s transform as follows:

$$X = W^{1/2}X \tag{2}$$

where $W_i(u_i,v_i)$ is an $n$ by $n$ matrix:

$$W_i(u_i,v_i) = \begin{pmatrix} w_{i1} & 0 & \cdots & 0 \\ 0 & w_{i2} & \cdots & 0 \\ \cdots & \cdots & \ddots & \vdots \\ 0 & 0 & \cdots & w_{in} \end{pmatrix} \tag{3}$$

In locally weighted regression models, the values of $W_i(u_i,v_i)$ are constant. In the GWLR model, on the other hand, $W_i(u_i,v_i)$ varies with the location i depending on the distance between location i and its neighboring locations (see eq. 4). The above process was repeated for each observation in the data, and consequently, a set of parameter estimates was obtained for each location.

The weight $W_{ij}$ in the weight matrix $W_i(u_i,v_i)$ is a decreasing function of distance $d_{ij}$ between the subject $i$ and its neighboring location $j$. Using the other spatial weighting function in GWLR model, we may produce a sub-sample with values of all zeros or ones. We need to employ the entire sample which makes this method more computationally intense. Thus, the spatial weighting function will be taken as the exponential distance-decay form:

$$w_{ij} = b_1\left(\frac{-d_{ij}{}^2{}_{ij}}{h^2}\right) \tag{4}$$

where $h$ is the kernel bandwidth. If the locations i and j coincide (i.e., $d_{ij}=0$), $w_{ij}$ equals one, while $w_{ij}$ decreases according to a exponential form as the distance $d_{ij}$ increases. However, the weights are nonzero for all data points, no matter how far they are from the center $i$ (Fotheringham et al. 2002).

The issue here is that solving for logit estimates is a computationally intense operation involving non-linear optimization, so a random sampling needs to be taken in large land cover change data set. Another problem is that non-parametric estimates based on a distance weighted sub-sample of observations may suffer from "weak data" problems. The effective number of observations used to produce estimates for some points in space may be very small. It can result in situations where the matrix X*X is non-invertible. This problem can be solved by incorporating subjective prior information during estimation.

## 2.2 *Temporal GWLR*

There will be two aspects to consider the temporal impacts in GWLR on land use change analysis. Firstly, different time intervals in different land cover change process on one point will cause different results. Secondly, the other kind of time intervals is the time interval between the land cover changes. In this study, for the first kind of temporal interval, we try to add an age variable to the regression model. For the second, we proposed a new weighting function that combines the "geographical space" and "temporal space" between one observation and its neighbors. We propose to modify eq. 4 as follows:

$$d^{*}_{ij}{}^2 = (u_i - u_j)^2 + (v_i - v_j)^2 + \lambda^*(t_i - t_j)^2 \tag{5}$$

where $t_i$ and $t_j$ are the time coordinates to different land use change periods. Since the spatial coordinate and temporal coordinate have different scale, all the coordinate data should be standardized. $\lambda$ is a temporal factor. The $d^*_{ij}$ will be adopted to compute the spatial-temporal distance in the following TGWLR.

TGWLR:

$$
\begin{aligned}
&y_i(u,v,t) \\
&= \log_e\left(\frac{p}{1-p}\right) = a(u,v,t) + b_1(u,v,t)x_{i1} + b_2(u,v,t)x_{i2} + \cdots + b_p(u,v,t)x_{ip} + b_{p+1}(u,v,t)age \\
&w_{ij} = \exp\left(\frac{-{d^*_{ij}}^2}{h^2}\right)
\end{aligned} \tag{6}
$$

Here, Nelder-Mead Simplex Method is employed to get $\lambda$ and bandwidth which make the model achieve the minimum residuals. This is a direct search method that does not use numerical or analytic gradients. If $n$ is the length of $x$, a simplex in n-dimensional space is characterized by the $n + 1$ distinct vectors that are its vertices. In two-space, a simplex is a triangle. At each step of the search, a new point in or near the current simplex is generated. The function value at the new point is compared with the function's values at the vertices of the simplex and, usually, one of the vertices is replaced by the new point, giving a new simplex. This step is repeated until the diameter of the simplex is less than the specified tolerance.

## 3 STUDY AREA AND DATA PREPARATION

### 3.1 *Study area*

In this study, part of the Calgary will be selected to the implementation. Seen from Figure 1, the study area is in the northeast of Calgary. The land cover changed dramatically in this region in the study period.

### 3.2 *Data preparation*

#### 3.2.1 *Data description*

The data used in this study, including sequential land cover data, demographic data, and transportation data, was obtained from three data sources: 1) land cover information derived

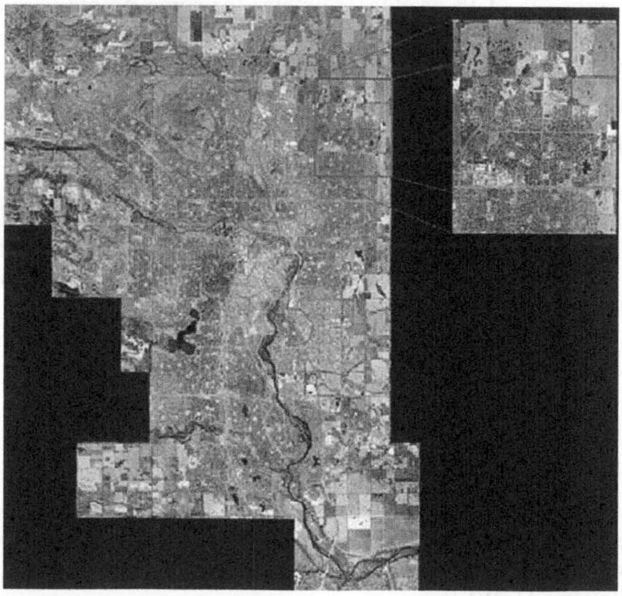

Figure 1. Image of calgary.

from digital orthophotos, 2) Census of Population data and 3) transportation data. The rural-urban land conversion data is using 1985–1990, 1990–1992, 1990–1999, 1999–2000 and 2000–2001 land cover changes.

Land cover data were obtained from the University of Calgary. All land cover files were rasterized at a resolution of 50 × 50 meter. Seven different types of land cover

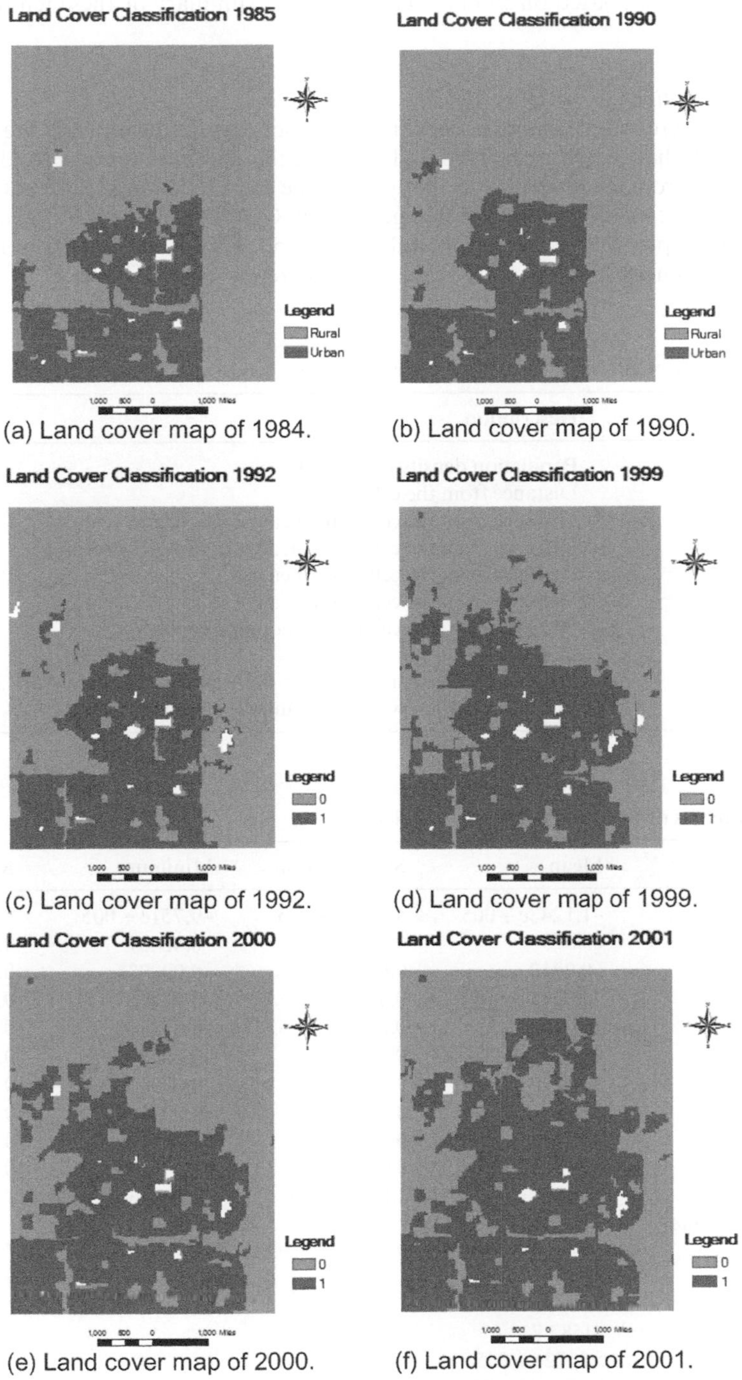


(a) Land cover map of 1984. (b) Land cover map of 1990. (c) Land cover map of 1992. (d) Land cover map of 1999. (e) Land cover map of 2000. (f) Land cover map of 2001.

Figure 2. Land cover maps of 1985, 1990, 1992, 1999, 2000 and 2001.

classification were employed: residential, commercial, industrial, Transportation, parks, vacant area, water bodies. The first four types are considered to be urban areas, parks and vacant area land cover is considered to be rural area that has potential of urbanization, and the last type are considered to be parcels that are not suitable for development (Fig. 2).

Inspired by other scholars, nine predictor variables were compiled in ArcInfo 9.1 via the spatial analyst module based on 50 m × 50 m cell size. A summary of these ten predictors is shown in Table 1.

### 3.2.2 *Data sampling*

To make sure there will be enough ones in the model estimation process. We first select the point which changed between 1985 and 2001. In the selected points, a random sample method samples from the spatial space firstly. Then the same location in different period will be selected to construct the land cover change data set. Considering the heavy computational intensity, 200 samples will be taken from different period. Thus, total 1000 samples are sampled in this test. The spatial locations of the samples are shown in Table 2.

Table 1. Summary of predictor variables for the rural-urban land conversion model.

| Variable name | Description |
|---|---|
| Pop_Dens | Population density of the cell |
| Dist_amenity | Distance from the cell to the |
| Dist_citycen | Distance from the cell to the nearest city centre |
| Dist_commserv | Distance from the cell to the nearest commercial service |
| Dist_lrtsta | Distance from the cell to the nearest |
| Dist_road | Distance from the cell to the nearest road |
| Dist_shopping | Distance from the cell to the shopping center |
| Slope | Slop of the cell |
| Per_avail | Percentage of urban land cover in the surround area within 200 m radius |
| Age | The temporal interval of the land cover change occur |

Table 2. Estimates of temporal GWR logit model.

| | Mean | Std. deviation | Minimum | Maximum |
|---|---|---|---|---|
| Constant | −1.1243e + 005 | 4.0885e + 005 | −9.751e + 005 | 3.4195e + 005 |
| Pop_Dens | 0.0853 | 0.36975 | −0.45401 | 0.84663 |
| Dist_amenity | −0.0012 | 0.0036722 | −0.008034 | 0.0050299 |
| Dist_citycen | −0.0071 | 0.013173 | −0.03345 | 0.012956 |
| Dist_commserv | −0.0016 | 0.0044668 | −0.011256 | 0.0033753 |
| Dist_lrtsta | 0.0042 | 0.0078722 | −0.008847 | 0.018968 |
| Dist_road | −0.0023 | 0.0036524 | −0.01063 | 0.0023831 |
| Dist_shopping | 0.0019 | 0.0062616 | −0.0061053 | 0.014764 |
| Slope | 0.1978 | 0.21486 | 0.028749 | 0.71012 |
| Per_avail | 7.8071e − 004 | 0.0020532 | −0.0013031 | 0.0049441 |
| Age | 374.8813 | 1.065e + 005 | −1.7096e + 005 | 1.95e + 005 |
| The sum of absolute residuals | 141.5164 | | | |
| Bandwidth | 61.404 | | | |
| $\lambda$ | 30.0238 | | | |
| Estrella R-squared | −1582.7 | | | |
| −Log likelihood | 5.5672e + 005 | | | |

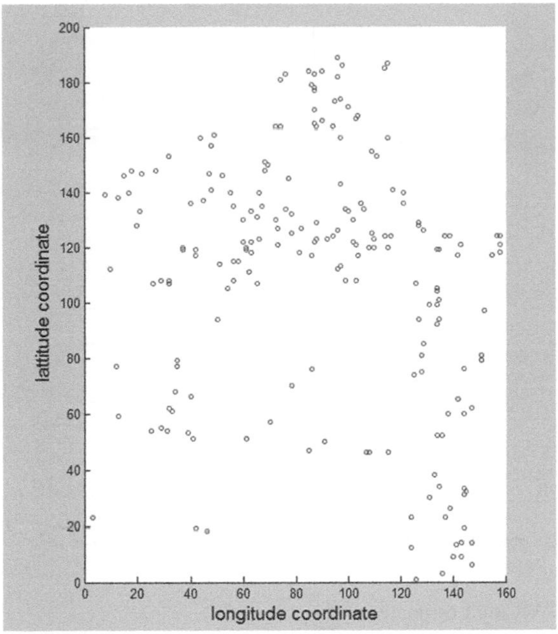


Figure 3. Map of sampling points.

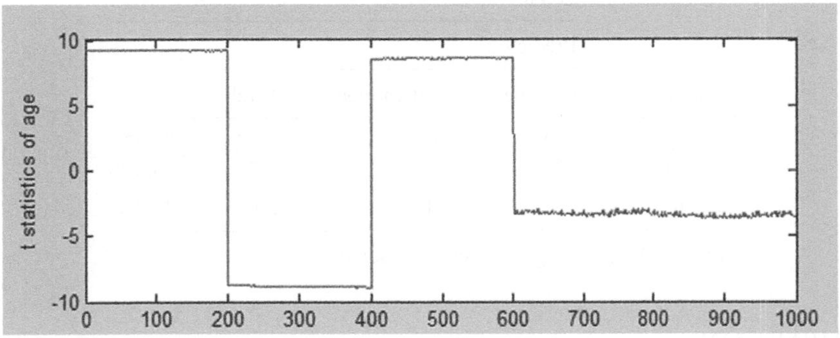


Figure 4. t statistics of age.

## 4 RESULTS AND DISCUSSION

Using the data and the equation (6), the Temporal GWLR model is carried out and the estimates are reported in Table 2, Figures 4 and 5 (every 200 points in the same period). The temporal factor $\lambda$ is 30.0238. And also Figure 4 shows the t statistics of age indicates age. All of the t statistics of age is larger than 3. So age is a very significant predictor variable. Both of those indicate temporal effect is strong.

Seen from the Figure 5, the coefficients have a remarkable drift in the temporal space. And in one period, the coefficients are almost the same. It indicates that there is temporal nonstationarity existing in land use change model and the spatial nonstationarity is not significant in such a small area. The prediction accuracy is shown in Table 3. The PCP (percentage correctly predicted) is 90.2%. And it has a good balance between change point and not change point.

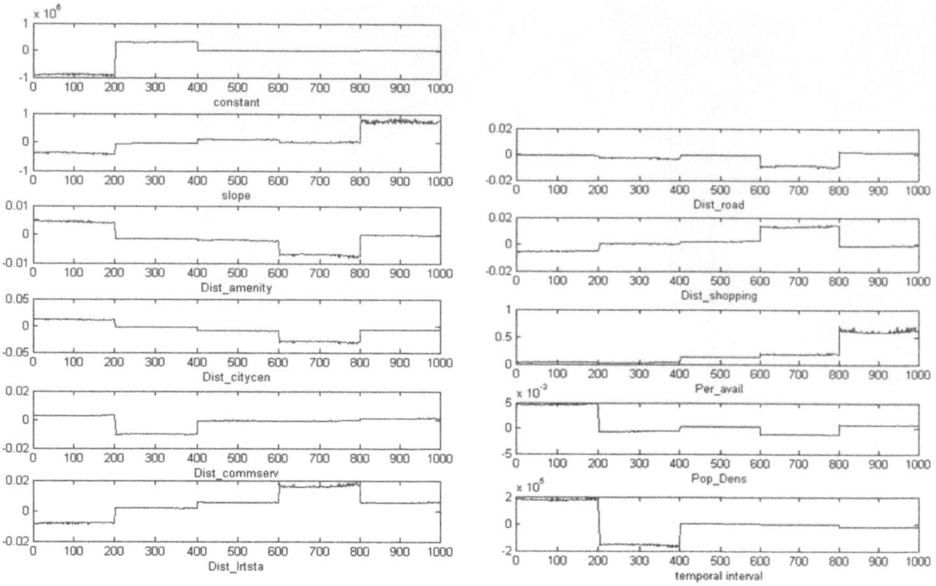


Figure 5. Temporal GWR logit estimates.

Table 3. Prediction accuracy of temporal GWR logit model.

| | GWR | | | |
|---|---|---|---|---|
| | Predicted | | | |
| Observed | Change | Not change | Total | PCP |
| Change | 160 | 62 | 222 | 72.07% |
| Not change | 36 | 742 | 778 | 95.37% |
| Overall | 196 | 804 | 1000 | 90.2% |

Note: PCP: percentage correctly predicted.

## 5 COMPARISON

### 5.1 *Logit model*

Using the same data and the following equation, the Logit model is carried out and the prediction accuracy are reported in Table 4 (77.6% vs. 90.2%). The sum of absolute residuals is 307.2313. Compared with it, obviously, the temporal GWR logit performs well.

$$y = \log_e\left(\frac{p}{1-p}\right) = \beta_0 + \beta_1 x_1 + \beta_2 x_2 + \cdots + \beta_K x_K \tag{7}$$

### 5.2 *Panel data model*

As the data set has both cross-sectional and time series dimensions, the application of panel data model with fixed effect is an appropriate choice for comparison. Panel data, in fact, allows for individual differences across individuals, or at least the unobservable and unmeasurable part of these differences, by modelling the individual specific effect. For comprehensive treatments see Hsiao (2003), Baltagi (2001), and Wooldridge (2002).

Table 4. Prediction accuracy of logit model.

| | Logit model | | | |
|---|---|---|---|---|
| | Predicted | | | |
| Observed | Change | Not change | Total | PCP |
| Change | 6 | 216 | 222 | 2.7% |
| Not change | 8 | 770 | 778 | 98.97% |
| Overall | 14 | 986 | 1000 | 77.6% |

Table 5. Prediction accuracy of panel data model.

| | Panel data with fixed effect | | | |
|---|---|---|---|---|
| | Predicted | | | |
| Observed | Change | Not change | Total | PCP |
| Change | 45 | 177 | 222 | 0.2027 |
| Not change | 31 | 747 | 778 | 0.9602 |
| Overall | 76 | 924 | 1000 | 79.2% |

Obviously, the temporal GWLR performs well.

Table 6. Prediction accuracy of GWR logit model.

| | GWR Logit model | | | |
|---|---|---|---|---|
| | Predicted | | | |
| Observed | Change | Not change | Total | PCP |
| Change | 8 | 214 | 222 | 0036 |
| Not change | 11 | 767 | 778 | 0.9859 |
| Overall | 19 | 981 | 1000 | 77.5% |

More formally, the panel version of the panel data model used here can be expressed in the following way:

$$y_{it} = \log_e\left(\frac{p_{it}}{1-p_{it}}\right) = X_{it}^{l}\beta + a + u_{it}, i = 1,2,\ldots,N, t = 1,2,\ldots,T \tag{8}$$
$$u_{it} = \mu_i + v_{it}$$

with $i = 1,2,\ldots,N$ denoting individuals, and $t = 1,2,\ldots,T$, denoting time periods. Where $\mu_i$ represent the individual differences. The sum of absolute residuals is 348.2928. The prediction accuracy is shown in the Table 5.

### 5.3 *GWR logit model*

Now we omit the temporal coordinates and take the equation 1 to modeling. The bandwidth is 7.4001. The sum of absolute residuals is 306.0643. The prediction accuracy is shown in the Table 6. The temporal GWLR model shows a better result.

Table 7. Total prediction accuracy of logit model in each period.

| | Logit model in each period | | | |
|---|---|---|---|---|
| | Predicted | | | |
| Observed | Change | Not change | Total | PCP |
| Change | 152 | 70 | 222 | 0.6847 |
| Not change | 32 | 746 | 778 | 0.9589 |
| Overall | 184 | 816 | 1000 | 89.8% |

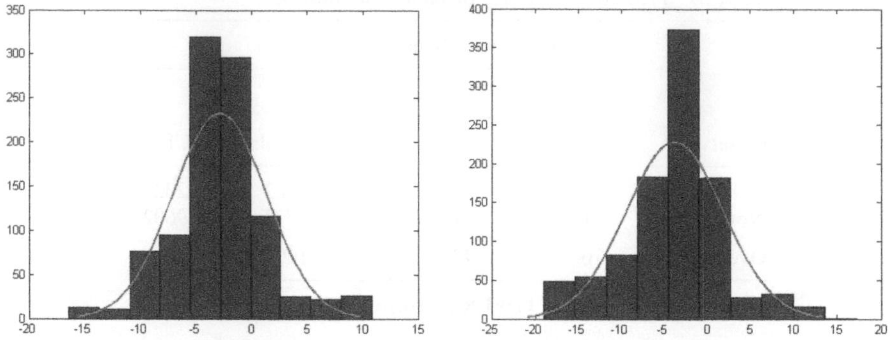


Figure 6. Histogram with superimposed normal density.

### 5.4 *Logit model in each period*

Sometimes, the models built in each period can partly solving the temporal nonstationarity. But it can't make good use the information from the other periods. So we try to build logit model in each period to compare with Temporal GWLR model. As the bandwidth of GWR logit model in each period in this data set is very large, the results of GWLR model are the same as logit model. Thus only the Logit Model in each period will be adopted for comparison here.

The PCP table shows there is only a bit difference between Logit Model in each period (89.8%) and temporal GWR model (90.2%). And the sums of absolute residuals (150.3476) of them are also very close to the one of temporal GWLR model (141.5164). In order to consider the difference between the estimations from those two models, the analysis of variance (ANOVA) will be performance. Also in order to explore the explanatory power of the model, Wilcoxon rank sum test will be adopted on the residuals from two models.

As ANOVA is based on the data which obey to the normal distribution, a transformation based on $\log_e(p/1-p)$ should be taken on the probability data. See from the Figure 6, we can take the transformed data from an approximate normal distribution.

From Figure 7 and Table 8, the p value is almost approach 0. ANOVA test indicates that there is a significant difference between the estimation of those two models.

The Wilcoxon rank sum test is a non-parametric statistical hypothesis test for the case of two related sample. It does not require assumptions about the form of the distribution of the measurements. Wilcoxon rank sum test for equal medians. It performs a two-sided rank sum test of the hypothesis that two independent samples, come from distributions with equal medians, and returns the p-value from the test. P is the probability of observing the given result, or one more extreme, by chance if the null hypothesis ("medians are equal") is true. Small values of P cast doubt on the validity of the null hypothesis. Here we get the p which is equal to 0.0048. So the difference between the residuals from the two models is significant. So, the temporal GWLR has significant improvement over the logit model in each period.

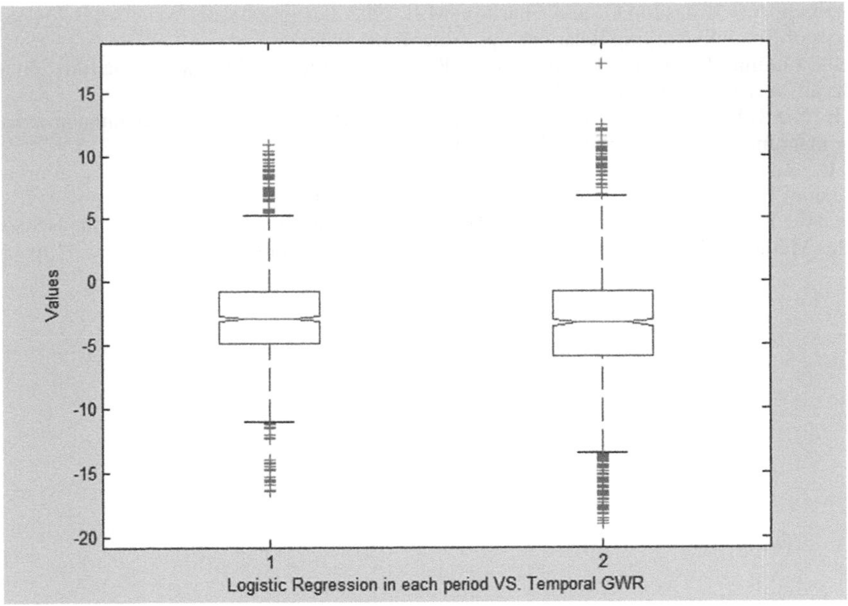


Figure 7. ANOVA test.

Table 8. ANOVA test logit in each period against temporal GWR.

| Source | SS | DF | MS | F | Prob > F |
|---|---|---|---|---|---|
| Columns | 437.5 | 1 | 437.548 | 17.42 | 3.12748e – 005 |
| Error | 50189.1 | 1998 | 25.12 | | |
| Total | 50626.6 | 1999 | | | |

SS = sum of squares; DF = degree of freedoms; MS = residual mean square.

## 6 CONCLUSIONS

In this paper, we model the land use change by Temporal GWLR model, and found some interesting profiles for the analysis of land use change in Calgary, Canada. Compared with the other models, Temporal GWLR model shows a significant improvement in the percentage of correctly predicted and the sum of absolute residuals. The results indicate that (1) significant spatial nonstationarity is not present in the data set; (2) temporal effect does exist and is significant.

However, several issues remain to be examined in future studies. Such as, the massive computational task limits the sample numbers in this model. There exists a possibility that unobservable factors (not covered by our data set) could influence the estimation results. In addition, the spatio-temporal dependence also should be considered in a spatio-temporal framework which is more flexible. Thus, a more elaborate version of Temporal GWLR model should be investigated. Nevertheless, our method represent a promising step for more elaborate econometric models to land use change modeling in a spatio-temporal framework.

## REFERENCES

Anselin, L. and A. Bera. 1998. "Spatial Dependence in Linear Regression Models with an Introduction to Spatial Econometrics." In A. Ullah and D. Giles (eds.), Handbook of Applied Economic Statistics. New York: Marcel Dekker, 237–289.

Fotheringham, A.S, Brunsdon C., and Charlton, M.E. 2002. Geographically Weighted Regression: The Analysis of Spatially Varying Relationships. Chichester: Wiley.
Hanjin Shi, Lianjun Zhang, and Jianguo Liu. 2006. A new spatial-attribute weighting function for geographically weighted regression. NRC press.
Huang, B., Xie, C.L., and Tay, R. (forthcoming2). Land use change modeling using unbalanced support vector machines. Environment and Planning B.
Huang, B., Zhang, L., and Wu, B. (forthcoming1). Spatio-temporal analysis of rural-urban land conversion. International Journal of Geographical Information Science.
Lagarias, J.C., J.A. Reeds, M.H. Wright, and P.E. Wright, "Convergence Properties of the Nelder-Mead Simplex Method in Low Dimensions" SIAM Journal of Optimization, Vol. 9, Number 1, pp. 112–147, 1998.

*Information Systems and Computing Technology – Zhang & Gu (eds)*
*© 2013 Taylor & Francis Group, London, ISBN 978-1-138-00115-2*

# One shot learning human actions recognition using key poses

W.H. Zou
*Faculty of Engineering, Tottori University, Tottori, Japan*
*Shanghai Institute of Technical Physics, Chinese Academy of Sciences, Shanghai, China*

S.G. Li
*Faculty of Engineering, Tottori University, Tottori, Japan*

Z. Lei & N. Dai
*Shanghai Institute of Technical Physics, Chinese Academy of Sciences, Shanghai, China*

ABSTRACT: In this paper, we focus on using depth data to classify short human actions. Every short action expresses a single meaning. We present a simple and effective approach which extracts "key pose" from the actions and focuses on the information centralized part of "key pose". With one shot learning, we model one kind of action as a series of key poses which are discriminable from the ones of other kinds of actions. Firstly, extract relatively static gesture from the actions, which are thought to contain a lot of discriminable information. This information may be what human hope to express. Secondly, use an algorithm similar with k-medoids clustering algorithm to find an "original gesture", which we regard as a situation human make no pose. Then use the "original gesture" to make the static gestures more discriminable to form the "key pose" of every action. Finally, process the actions need to be recognized in the same way and then match the corresponding "key pose", which is realized by a point cloud probability method. Compared to the past studies, this novel method is simple and quick. And our experiment results based on some depth datasets from the ChaLearn Gesture Dataset show that this method is effective for a large proportion of these tasks.

## 1 INTRODUCTION

### 1.1 *Background*

The recognition of human actions is a fundamental and very important branch of computer vision. Until now, people have made many efforts to develop all kinds of methods for learning and recognizing human actions by machine. It can be very useful in a wide range of areas. For example, gesture language exchange, human-computer interaction, game development, robot technology and so on.

Researchers have developed many difference ways to express and recognize the human actions. For example, A. Fathi and G. Mori use the motion-based method. M. Blank and Y. Ke regard actions as space-time shapes. Also J.C. Niebles and C. Schuldt regard actions as space-time interest points. K. Schindler use action snippets to express and recognize the actions. In this paper, we will express a short action containing a single particularized implicature with several "key poses" which have been focused both spatially and temporally. It's a novel representation method for action feature extracting and action recognition.

There are many kinds of different method to detect, track and recognize human body motion from video. It has been studied for more than twenty years. But it's still full of challenge to recognize continuous and natural body motion due to the high deformability and high level of self-occlusion. It's discussed by V. Ganapathi. These problems can be partially solved by the real-time depth cameras according to the research of M. Siddiqui and C. Plagemann. The Kinect of Microsoft is an ideal depth camera to realize 3D human body recognition because its consumer price to be widely commercial used.

In this paper, we will use the depth video captured by the Kinect to realize 3D human motion recognition. The approaches for dealing with 3D shape can be roughly divided to two classes. First class based on a model. And the model is deformed at each time step of the sequences to obtain a globally consistent shape representation. So it's likely to lose some detail geometrical or texture information. For example, Juergen Gall relies on a skeletal model of the human body with 10 degrees-of-freedom for the joints and 6 additional parameters for the rotation and translation of the torso. Fit the model to the 3D shape then adjust the parameters frame by frame. The system does not make use of prior information about the objects but it relies on a skeletal model of the human body with 10 degrees-of-freedom. So when there is serious occlusion of the human skeleton, the performance of this method will be reduced. Even sometimes we may have only part of human body in our visual field. Second class is called model-free method. For example, Avinash Sharma first compute a set of initial sparse correspondences by matching features computed over any scalar function defined on the shape surface. Then use a multi-scale heat kernel descriptor to propagate initial correspondences over the shape.

Our approach is motivated by the video preview. When we choose videos, we will show a video preview of the video. The preview of video may be only one pose, which is relatively a representative frame among all the frames. On the sight of these key pose, Human can more or less classify these videos into several classes according to the similarity between these key poses. It's miraculous ability. And this method can be used to automatic human action recognition.

In contrast to past studies, our approach is a much more simplified model-free 3D human action recognition method, which is based on the "key pose". It's qualified for a lot of actions. And it can be used to classify human actions quickly and preliminarily.

### 1.2 *Related work*

Human action recognition has been studied extensively in recent years. But many approaches take a lot of time to train a large number of images. Recently, one shot learning human action recognition is proposed and becoming a popular branch of computer vision. One shot learning is broad-spectrum especially in interactive game, robot vision and human-computer interactive. But it's more difficult to achieve high accuracy and robustness. We strongly believe that "key pose" based approach is a suitable one shot learning human action recognition method for the actions with relatively discriminable poses.

As far as we know, "key pose" has been seldom used before for human action recognition. Y. Wang and G. Mori have proposed to recognize human actions from still images with latent poses. Different from other work that learns separate systems for pose estimation and action recognition, then combines them in an ad-hoc fashion, their system is trained in an integrated fashion that jointly considers poses and actions. They directly exploit the pose information for action recognition. And they have demonstrated that it's a good way to improve the final action recognition results by inferring the latent poses. Different from their work, we propose to extract "key pose" from training data and test data, then use the key pose to recognize human actions. Our algorithm is based on the underlying principles that action can be inferred from latent poses to a certain extent.

F.J. Lv and R. Nevatia proposed to model the 3D human actions as a series of synthetic 2D human poses rendered from a wide range of viewpoints. The synthetic poses are represented by a graph model called Action Net. The major difference between the Action Net and graph models such as Hidden Markov Models and Conditional Random Fields is that, information such as camera viewpoint and action connectivity is explicitly modeled in the Action Net, while these graph models use parameters to encode such information. They only extract key poses from model actions, then formulate action recognition as matching Action Net with every frame in the input sequence. Different from their work, we prepare the model action as discriminable key pose, and then prepare the input sequence as key poses ordered by specificity of every pose, which will be effective and cost less computation. On the other hand, we directly use the depth data from a 3D camera called Kinect instead of expressing 3D pose with a lot of 2D poses rendered from a wide range of viewpoints. They define key poses as the poses with maximum or minimum motion energy in a sliding window whose length is fixed to be L. With less omission ratio and randomness, we consider both duration and specificity to extract key poses.

The approach closest to ours is proposed by Sermetcan Baysal, Mehmet Can Kurt and Pinar Duygulu. Different from most previous studies which employ histogram to represent the pose information present in each frame and result in the loss of spatial information among the components forming the pose, they preserve and utilize spatial information encapsulated in poses. And oppositely temporal information is totally disregarded.

In this paper, we propose an approach to recognize action preliminarily and fast. Also it's a kind of one shot learning human action recognition. We firstly find some "static poses" in the time scale, which are supposed to be relatively static and contain some special information human want to express. These "static poses" are considered as candidates of "key poses". Our approach not only focuses on the spatial information but also take temporal information into account. Then we pick up key poses among the static poses of all the model actions instead of extracting common pose inside only one action by the k-medoids clustering algorithm. Different from previous works, we extract key pose for both training actions and test actions. By the way, as the advantages of the depth data, it's simpler and more effective to calculate similarity between poses and movement.

## 2 EXTRACT ORIGINAL POSE

### 2.1 *Dataset description*

There are several batches, which are provided by the CHALEARN gesture challenge. A batch is played by one person according to a certain script. Every batch includes about one hundred of actions, which are contained in 47 videos. These videos can be divided into two classes, model actions and test actions. Every model video includes one model action. A test video may contain one to five actions. As one shot learning action recognition, every model action can be watched only one time.

### 2.2 *Extract relatively static pose*

Need to be declared, there is a basic assumption. We suppose there is one kind of pose in every action, which is called original pose. Original pose is one kind of natural pose when human doesn't do any pose. Human always start from an original and natural pose, then do an action contain special meaning, and finally end the action and return to the natural pose. For example, when a person does an action for calling taxi, his two hands are thought to be firstly relaxed and fallen naturally. Then his hands are put up to form a pose for calling taxi. Finally he put down his hands and his hands will return to be relaxed, so hands fall naturally. Until now, the whole action is finished. In this action, the pose with hands naturally falling down is defined as "original pose".

If a pose and its similar pose appear in a certain continuous time, this pose will be defined as a relatively static pose. The poses which are discriminable enough to distinguish an action from the others are called "key pose". A relatively static pose can be candidates of original pose and key pose. So primarily, we should extract the relatively static pose from the video. It can be simply realized with the depth data.

Firstly, with the difference between continuous frames we can directly calculate the movement between two frames. Then we erode the difference image for several times using a simplest template, which is a 3 × 3 square. We check the result image. If all the movement hasn't been removed, the pose of this frame will be abandoned; else this pose will be taken into further observation. And if a pose last for more than a certain time $T$, this pose will be regarded as "static pose".

### 2.3 *Find original pose*

As one shot learning human action recognition, we have only one model action for every category of actions. We use the method mentioned above to pick up all the static poses from every model action. Then we put them all together to form a posture set. $\{g_1, g_2, ..., g_i, ... g_N\}$

We have supposed that action should start with natural state of human body and end with natural state of human body. And such natural states should be similar postures for one set of approximate actions. Other pose will more or less have some discrimination according to its action. So the occurrence frequency of original pose should be the highest among the postures set. The original pose we describe here is a collection of several similar poses which can be taken as original pose.

For a pose $g_i$ among the posture set $\{g_1, g_2, ..., g_i, ... g_N\}$, we define the occurrence frequency of poses, which are similar with $g_i$, as $F_i$:

$$F_i = \frac{1}{\sum_{j=1}^{N} similarity\ (g_i, g_j)} \tag{1}$$

The original pose $g_0$ should be the most frequent state of human body.

$$g_0 = g_{\max(Fi)} \tag{2}$$

The result of searching for original pose is shown in fig. 1. In fact, the original pose should be a collection of many similar poses, but we display it using one pose which is thought to be most expressive.

## 3 VIDEO SEGMENTATION

With the help of original pose extracted from every batch, the videos in the corresponding batch can be easily and effectively segmented to several sequences. Every sequence will

Figure 1. Original pose. Every pose here correspond to the original pose of one batch.

contain only one action. According to the feature of depth image, we simply calculate the difference image between original pose and the pose in every moment. The nonzero pixel in the difference image can be divided into three classes. First class show the fluctuation of a normal rang caused by undulate the human body or random errors of detector. Second are some extreme points caused by the detector defect or occlusion of human's body. Because of the occlusion of human's body, there may be some place where infrared light can't reach. Third kind of points shows the special information of the pose at that moment. We define the number of the third kind of points in the difference image of $i$th frame as $f_i$. As the fluctuation of human's body and random errors of detector are in a small range and the errors caused by body occlusion are always extreme points, it's easy to remove these points by threshold processing. Finally, erode the processed difference image with a 3 × 3 template to remove small fluctuation of body in the direction of image plan. Then count the number of the third kind of points of every frame in the sequence.

To segment a video into several actions, we slide a sliding window over the sequence. The length of the sliding window is $L$. If the variable $f_i$ of a frame is always the minimum inside such a sliding window, we will regard this frame as a demarcation point between actions. And if the variable $f_i$ of a frame is zero, it will directly be taken as demarcation point. The frame per second in the video is 10. We choose $L = 16$ frames in our system, because we suppose one action will last longer than such a short time. The results of segmentation are shown in fig. 2.

## 4 RECOGNIZE ACTIONS USING KEY POSE

We propose to recognize actions using key pose following these steps:

- Find key poses based on the relatively static poses and original poses for every action. Then focus on the moving part of human body inside every key pose, according to the amount of movement in a certain period of time.
- Calculate the similarity between two parts of human's body, which belong to two different key poses.
- Match the key pose of test action to the key poses of model actions.

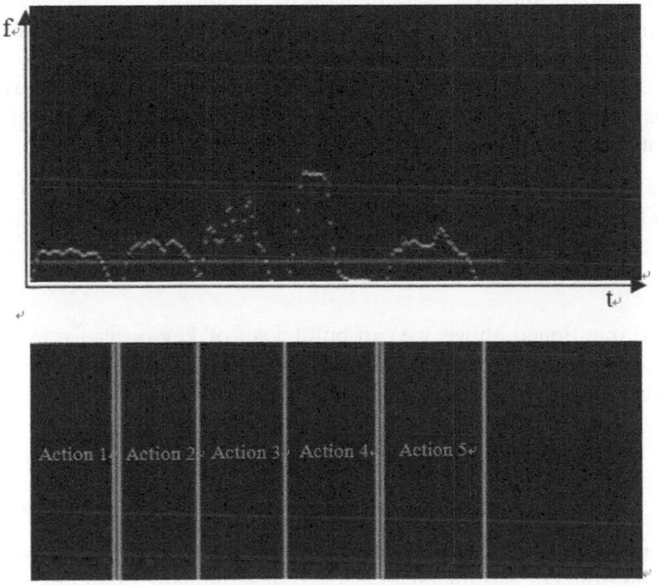


Figure 2. Action segmentation. The image up shows the variable fi of one sequence. Horizontal axis shows the frame number. Ordinate axis shows the variable f. The image down shows the segmentation result. Every action in the sequence is marked by green lines.

### 4.1 *Temporally focus on the key pose*

Key pose can be described as the ones, which are representatives in a specific action. In the section of finding relatively static poses, we have grouped the frames, which show common pose appearances inside one action. However in the section of finding original pose, we have grouped the frames, which show common pose appearances in different types of actions. Thus, we base our key pose extraction process on the static poses and original pose.

We suppose that we extract several sequences of static poses from an action, which has been segmented using the method mentioned above. Every sequence consists of several gestures. A sequence of static pose can be described as {gesture 1, gesture 2 ... gesture N}. The first frame, last frame and original pose are respectively defined as {gesture-begin, gesture-end, gesture- standard}. And if any gesture of one sequence is the same one with gesture-begin, gesture-end, or gesture-standard, this sequence will be removed.

The standard used to judge one gesture to be the same with another gesture is defined like this:

- Calculate the difference image of these two frames.
- Erode the difference image with a $3 \times 3$ template for three times.
- If the difference is completely eroded, these two poses will be regarded as the same one.

Some sequences of static poses will survive to the end. And every survived sequence corresponds to one key pose. Here, the middle frame of the sequence will roughly be used to represent the key pose. One result is shown in fig. 3 as an example.

### 4.2 *Spatially focus inside the key pose*

The process above is about finding key poses in action sequence, which can be regarded as temporal focusing. Next, we propose to spatially focus inside the key pose by finding the moving part of body at that moment, which will make the key pose more representative.

- Firstly, we pay attention to frame sequences a short time before and after the frame sequence of static poses which including a key pose. Calculate the image difference between two continuous frames in such sequences. After threshold processing, the reserved points in the difference image show the movement at the moment.
- Secondly, extract different parts of body between key pose and original pose, which are thought to be limbs that have been moved.
- Then make use of the two parts information above, and choose the part of body which contains most movement points. It's thought to be a partly pose which has appeared exactly at that moment.

We think such a partly pose will be more representative and easier to distinguish. The example can be found in fig. 4.

### 4.3 *Match key poses2*

With the method mentioned above, we can build a set of key poses for every model action, which can be used as model for key pose matching. Corresponding to the $i$th action, define a set of key poses:

$$g_j = \begin{cases} \{v_1, v_2, \ldots, v_k\}, 1 \le i \le N \\ g_0, i = 0 \end{cases} \tag{3}$$

Notice that, the number of model actions is $N$. $g_0$ is the representation gesture of the original pose. The range of parameter $k$: $k >= 0$. $k = 0$ means there are no key pose inside the action.

On the other hand, define a set of key poses corresponding to a certain action, which need to be recognized.

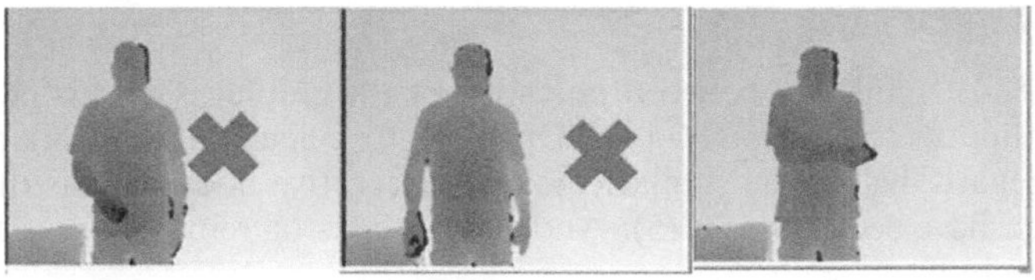

Figure 3. Key pose. There are three static poses in an action. After left one and middle one are removed, right pose is regarded as the key pose of this action.

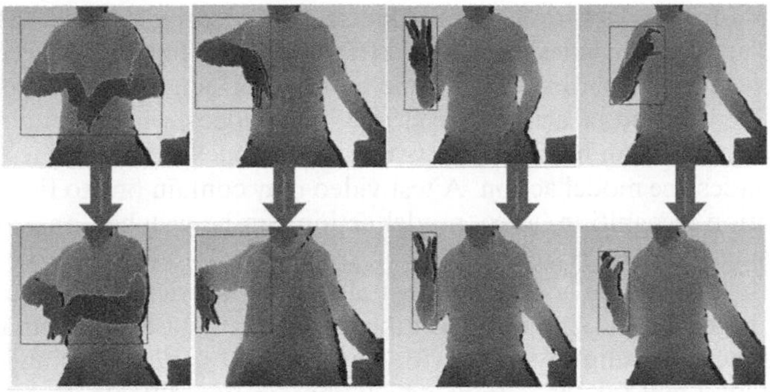

Figure 4. Matched key pose. The poses above are temporally and spatially focused poses from action need to recognize; the poses below are successfully matched key poses corresponding to the poses above.

$$G = \{v_1, v_2, \dots\} \tag{4}$$

Where, every key pose has been descending ordered by its specificity. The specificity of $i$th key pose inside the set $G$ is defined as:

$$Gv_i \cdot specificity = similarity_{pose-to-pose}(Gv_i, g_0) \times t_{Gv_i} \tag{5}$$

The first term is the similarity between $i$th key pose of the set $G$ and the original pose. The second one is the duration of $i$th key pose of the set $G$. It's equal to the length of the sequence that the key pose belongs to.

For matching a set of poses from test action to another set of poses from training action, we define the similarity between $i$th key pose inside the set $G$ and the set of poses corresponding to $j$th model action like this:

$$similarity_{pose-to-pose}(Gv_i, g_j) = \min_k \{similarity(Gv_i, g_{jv_k})\} \tag{6}$$

And the matching from one pose set of test action to another pose set of training action is judged by the similarity calculated by (6).

$$G \propto \min_{similarity(Gvi, gj)} g_j \tag{7}$$

For an action need to recognize, we extract a set of key poses. Firstly, we use the first pose $G(v_1)$ to find matched model action. As the poses in the set $G$ have been ordered according to their specificity, the first pose $G(v_1)$ is thought to be the most representative pose. If the pose $G_{vi}$ can't be successfully matched to one model action or the pose is matched to the original poses, we will resort to the next key pose $G(v_{i+1})$, until the end of the poses or successful matching.

### 4.4 *Similarity between poses*

As the estimation of similarity between poses is not the emphasis in this paper, we simply employ the outline matching method of OpenCV to calculate the similarity of outline sequences. We regard the outline similarity of representative body part as the similarity of two poses, which have been used in (5). And the outlines of representative body parts are drawn with green in fig. 4.

## 5 EXPERIMENT RESULTS

We demonstrate our approach on a public dataset provided by the CHALEARN gesture challenge. There are many batches played by different people. In a batch, according to a certain script, one person will perform about one hundred of actions, which belong to 8 to 15 action classes. One hundred of actions are stored in 47 videos, which contain 1 to 5 actions respectively. These videos can be divided to two classes, model actions and test actions. Every model video includes one model action. A test video may contain one to five actions. As one shot learning action recognition, every model action can be watched once only. The video segmentation using key poses is successful.

We find that the recognition accuracy of our algorithm is seriously impacted by the rough method of calculating outline similarity. But the test on the dataset shows that our approach which recognizes actions using key poses with temporal and spatial focusing is especially and relatively performing well for the actions which contain tiny difference. And these actions should have been more difficult to recognize by other approaches.

As I. Guyon and *Di Wu* described in their papers, the third and tenth batches of develop data in CHALEARN gesture challenge 1 are especially difficult to distinguish, respectively with an error rate 0.5 and 0.6, which have been calculated as a Levenshtein distance of the recognition results. The recognition result for some of these actions is shown in table 2. Every action is represented by its key pose in fig. 5.

In the experiment, we found that with our approach we can get really a good representation for most actions. Based on the assumption that hesitating for a short time is very common in human actions and also it's necessary to express some special meaning sometimes.

Table 1. Segmentation.

| Batch number of the data | Segmentation mistakes rate % |
|---|---|
| 1 | 0 |
| 2 | 3 |
| 3 | 2 |
| 5 | 1 |
| 7 | 1 |
| 10 | 0 |

Table 2. Accuracy for extracting key poses and recognize.

| Action number | Accuracy of extracting Key pose correctly % | Recognition accuracy % |
|---|---|---|
| 1 | 100 | 88.9 |
| 2 | 90.9 | 62.7 |
| 3 | 100 | 36.4 |
| 4 | 100 | 77.8 |

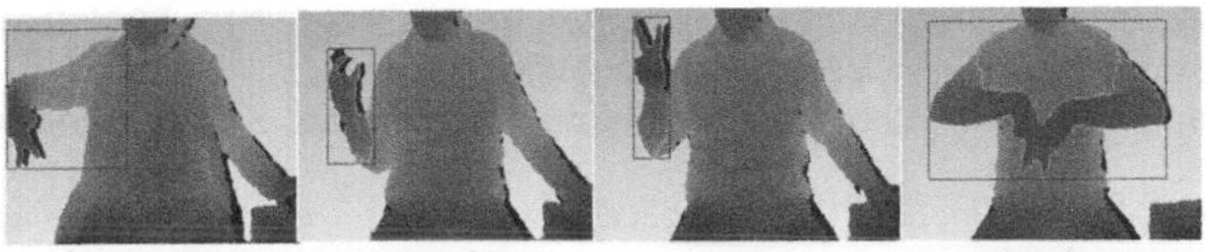


Figure 5. Actions represented by key pose.

Using key poses as the representation of actions help us to recognize actions primarily and quickly. Also it has special advantages for distinguishing the actions which are slightly different from each other. It owes to the feature of our approach that it will automatically focus on the special movement inside key pose.

However, there are still some problems with the approach. As shown in the table above, though extract key poses exactly from the actions, it's difficult to match the key pose to a right one. It's due to the rough calculating method for similarity between poses.

## 6 CONCLUSIONS

In this study, we discuss a new method for representing human actions then use it to recognize human actions after one shot learning. Moreover, we spatially focus on the movement inside key pose, which can be realized with a simple method benefit by the characters of depth data. It's different from previous studies, which only use the pose information with histograms or oppositely preserving the spatial information but completely omit the temporal information. But our method for calculating similarity between poses is so rough to seriously impact the recognition result. In the future works, we may focus on developing a method suitable for calculating the pose's similarity in our condition and further improve the recognition result.

## REFERENCES

Baysal, S. & Kurt, M.C. 2010. "Recognizing Human Actions Using Key Poses," ICPR, pp. 1727–1730.

Blank, M. & Gorelick, L. 2005. "Actions as space-time shapes," ICCV, vol.2, pp. 1395–1402.

Dorner, B. 1993. 1. "Hand shape identification and tracking for sign language interpretation," In IJCAI Workshop on Looking at People.

Dollar, P. & Rabaud, V. 2005. "Behavior recognition via sparse spatio-temporal features," VS-PETS, pp. 65–72.

Efros, A.A. & Berg, A.C. 2003. "Recognizing action at a distance," ICCV, vol.2, pp. 726–733.

Fathi, A. & Mori, G. 2008. "Action recognition by learning mid-level motion features," CVPR, pp. 1–8.

Gall, J. & Fossati, A. 2011. "Functional Categorization of Objects using Real-time Markerless Motion Capture," CVPR, pp. 1969–1976.

Guyon, I. & Athitsos, V. 2012. "ChaLearn Gesture Challenge: Design and First Results," CVPRW, pp. 1–6.

Ganapathi, V. & Plagemann, C. 2010. "Real time motion capture using a single time-of-flight camera," CVPR, pp. 755–762.

Ke, Y. & Sukthankar, R. 2007. "Spatio-temporal shape and flow correlation for action recognition," CVPR, pp. 1–8.

Lv, F.J. & Nevatia, R. 2007. "Single View Human Action Recognition using Key Pose Matching and Viterbi Path Searching," CVPR, pp. 1–8.

Niebles, J.C. & Wang, H. 2008. "Unsupervised learning of human action categories using spatial-temporal words," Int'l J. Computer Vision, 79(3). 299–318.

Nowozin, S. & Bakir, G. 2007. "Discriminative subsequence mining for action classification," ICCV, pp. 1–8.

Plagemann, C. & Ganapathi, V. 2010. "Real-time identification and localization of body parts from depth images," ICRA, pp. 3108–3113.

Sharma, A. & Horaud, R.P. 2011. "Pologically-Robust 3D Shape Matching Based on Diffusion Geometry and Seed Growing," CVPR, pp. 2481–2488.
Schuldt, C. & Laptev, I. 2004. "Recognizing human actions: a local SVM approach," ICPR, vol. 3, pp. 32–36.
Schindler, K. & Gool, L.V. 2008. "Action snippets: how many frames does human action recognition require?" CVPR, pp. 1–8.
Siddiqui, M. & Medioni, G. 2010. "Human pose estimation from a single view point, real-time range sensor," CVPRW, pp. 1–8.
Wang, Y. & Sabzmeydani, P. 2007. "Semi-latent Dirichlet allocation: a hierarchical model for human action recognition," ICCV, Workshop on Human Motion.
Wang, Y. & Mori, G. 2010. "Recognizing Human Actions from Still Images with Latent Poses," CVPR, pp. 2030–2037.
Wu, D. & Zhu, F. 2012. "One Shot Learning Gesture Recognition from RGBD Images," CVPRW, pp. 7–12.
Zhu, Y. & Fujimura, K. 2007. "Constrained optimization for human pose estimation from depth sequences," ACCV.

*Information Systems and Computing Technology – Zhang & Gu (eds)*
 *ISBN 978-1-138-00115-2*

# Band grouping pansharpening for WorldView-2 satellite images

X. Li
*Jiangsu R&D Center for Internet of Things, Wuxi, Jiangsu, P.R. China*

ABSTRACT: Pansharpening is an important tool in remote sensing applications. With the availability of very high resolution remote sensing imagery, more and more challenges come to the research field. In this paper, a novel pansharpening method is proposed. First, original multispectral bands are grouped according to the relative spectral response between the multispectral and panchromatic sensors. Then the grouped bands are processed through different sharpening models. The experimental evaluations are carried out on WorldView-2 data. Visual and objective analyses show that the proposed band grouping strategy works well and outperforms some existing methods.

## 1 INTRODUCTION

Pansharpening, a branch of image fusion, is a key issue in many remote sensing applications requiring both high spatial and high spectral resolution. The technique may be defined as the process of synthesizing multispectral images at higher spatial resolution, and is receiving ever increasing attention from the remote sensing field. Many spaceborne imaging sensors, which operate in a variety of spectral bands and ground scales, can provide huge volumes of data having complementary spatial and spectral resolutions. Due to the constraints on the signal to noise ratio, if the requested spectral resolution is higher, the spatial resolution must be lower. Usually, the panchromatic (PAN) and multispectral (MS) images acquired by satellites are not at the same resolution. MS images have a high spectral and low spatial resolution while PAN image has a high spatial resolution without spectral diversity. Pansharpening specifically aims at increasing the spatial resolution of MS images as well as preserving their spectral signature by using the spatial information of PAN image. A wide variety of pansharpening techniques have been developed during the last two decades, and the pansharpened products are highly desirable in many applications, such as classification, mapping, detection, Google Earth, and so on. (Camps-Valls et al. 2011, Longbotham et al. 2012).

A good pansharpening approach should have at least two essential components, i.e. enhancing high spatial resolution and reducing color distortion. Initial efforts based on Component Substitution (CS) methods, like the Gram-Schmidt (Laben & Brower 2000), Brovey Transform (BT) (Liu 2000), principal component analysis (Wang et al. 2005) and Intensity-Hue-Saturation (IHS) (Tu et al. 2001), are mainly focused on enhancing spatial resolution for easing tasks of human photo-interpretation. To overcome the color distortion in CS-based pansharpening process, other researchers developed algorithms based on multiresolution analysis, like Laplacian pyramid (Aiazzi et al. 2002), discrete wavelet transform (Zhang 1999), curvelet transform (Choi et al. 2005), or contourlet transform (Mahyari & Yazdi 2011), which extract zero-mean high-pass spatial information from PAN image and inject them into MS data. However, in the process of high-pass details injection, spatial distortions, typically, ringing or aliasing effects, originating shifts or blur of contours and textures, may be induced, or the spatial resolution of MS images may not be sufficiently enhanced.

Recently, the new-style very high resolution satellite imagery presents different features and brings more challenges. WorldView-2 (WV-2) satellite, launched by DigitalGlobe on

Oct. 8, 2009, is the first commercial satellite to carry a very high spatial resolution sensor with one PAN and eight MS bands. As shown in Figure 1, the MS bands (Band1 = Coastal, Band2 = Blue, Band3 = Green, Band4 = Yellow, Band5 = Red, Band6 = Red Edge, Band7 = NIR1, Band8 = NIR2) cover the spectral range from 400 nm–1050 nm at a spatial resolution of 1.84 m, while the PAN band covers the spectrum from 450 nm–800 nm with 4× greater spatial resolution, 0.46 m. Compared to IKONOS or QuickBird imagery having four MS bands (Blue, Green, Red, and NIR) and one PAN band, WV-2 data present some notable features, such as more MS bands, spectral coverage of PAN, and the relative spectral responses between PAN and MS bands. It can be seen from Figure 1 that both the Coastal and NIR2 bands are almost out of the PAN spectrum of WV-2. Moreover, the responses of Blue and Green bands have been improved to match that of PAN. These new features lead to the fact that most existing algorithms can hardly produce satisfactory pansharpened results. Although Padwick et al. (2010) proposed a Hyperspherical Color Sharpening (HCS) method for WV-2 imagery which is based on the hyperspherical color transform and handles all 8-band MS data at one stroke and produced acceptable visual results, evident color distortions still happen in this scheme. Recently, Tu et al. (2012) developed an Adjustable Pansharpening approach (APS) trying to cope with some high-resolution in-orbit spaceborne imaging sensors. They investigated the intensity match problem of the different sensors and integrated IHS, BT, and SFIM methods. The developed APS approach shifts among five different modules by two tunable parameters. To improve spectral fidelity of WV-2 image fusion, Ji et al. (2012) designed a constrained generalized IHS model with localized weight structure through land cover classification of the MS data. However, the model needs more priori information of the researched area and depends partially on the operator's experience.

In this paper, a novel and effective pansharpening method for WV-2 satellite images is proposed, whose core consists of: (1) it conducts a relative spectral response based grouping strategy to the input MS bands; (2) two different pansharpening models are developed to process the grouped data. To verify the efficiency of the proposed strategy, experimental evaluations are carried out on true WV-2 data set. The visual and quantitative analyses of the different pansharpened results prove that the proposed method performs well and can achieve high spatial and spectral quality.

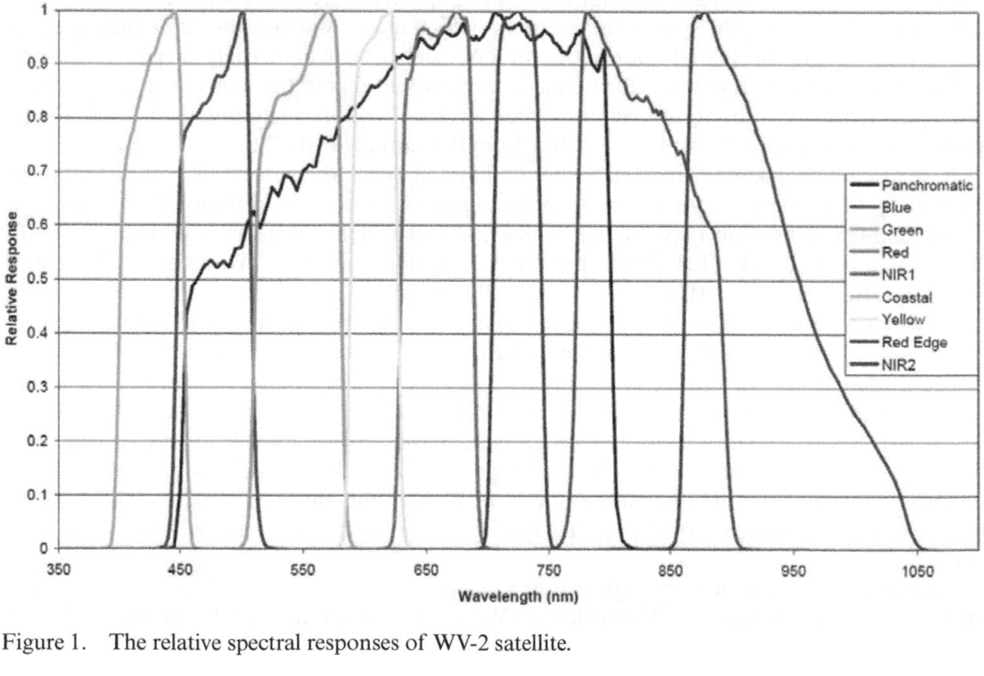


Figure 1. The relative spectral responses of WV-2 satellite.

## 2 THE PROPOSED METHOD

### 2.1 *The band grouping strategy*

According to the actual relationship between spectral response of PAN and MS bands displayed in Figure 1, it is easy to find that (1) the Blue, Green, Yellow, Red and Red Edge bands are principally covered by the response of PAN; (2) the NIR1 band is only partially covered; and (3) the Coastal and NIR2 bands almost fall out of PAN's spectral curve. Thus, WV-2 8-bands MS data can be roughly divided into two groups: one has six bands including Blue, Green, Yellow, Red, Red Edge and NIR1, the other has Coastal and NIR2 which have little spectral relationship with PAN.

Suppose that the original PAN and 8-band MS images were acquired at the same time and were already co-registered. The proposed pansharpening scheme is shown in Figure 2. The original MS bands (Band1, ..., Band8) are expanded to the same size as the PAN image and the upscaled low resolution MS images are denoted by $LRM_1$, ..., $LRM_8$, respectively. Based on the grouping strategy, the input MS data are divided into two parts. $LRM_1$ and $LRM_8$ belong to one group and the other six bands to another group. Then, two different models (Model-1 and Model-2) employ PAN image to sharpen the two grouped MS images individually. Finally the pansharpened results, denoted by $HRM_1$, ..., $HRM_8$, are obtained.

### 2.2 *Model-1*

Model-1 is developed on the assumption that there is no spectral relationship between PAN and MS images. The spatial information of PAN image can be individually injected into $LRM_1$ and $LRM_8$. It is noted that Band1 and Band8 are resized by nearest neighbor interpolation and the data of $LRM_1$ and $LRM_8$ are preserved as in the original images exactly. Since the nearest neighbor process does not introduce any new data into the expanded MS image, the original spectral signature keeps very well. There are two steps to implement Model-1.

Step 1: A window-based injection mode is proposed. The spatial information of PAN image is multiplicatively injected into the MS image and the calculation can be expressed as

$$HM_i(i,j) = \mu_w(LRM_i) \times \frac{PAN(i,j)}{\mu_w(PAN)} \tag{1}$$

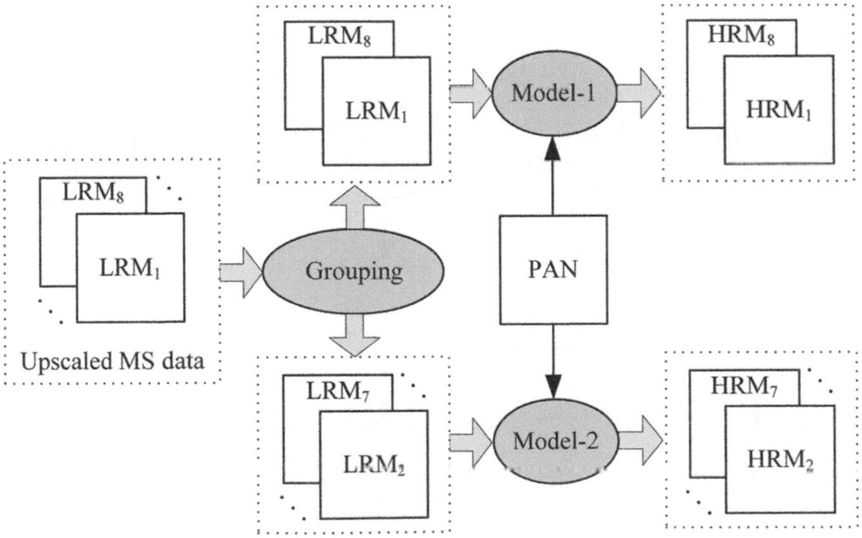


Figure 2. The proposed band grouping pansharpening scheme.

where $i = 1$ or 8, and $\mu_w$ is the average value in the window $w$. The window $w$ has the size of $4 \times 4$, and moves from the top-left corner to the bottom-right of the image without overlap. The new values for all the pixels in the sliding window are calculated.

Step 2: The redundant wavelet transform is employed to remove some block distortions which may occur on the edges of the high spatial resolution MS images ($HM_1$ and $HM_8$). Here "à trous" algorithm is used to minimize or remove them. First, the decomposition process is applied to both PAN and each MS image ($HM_1$ or $HM_8$), respectively. Next, the wavelet plane of each MS image is replaced by the wavelet plane of PAN. After the replacement, reconstruction process is conducted to each band to compose the final sharpened image ($HRM_i$).

### 2.3 *Model-2*

Model-2 is proposed as a Correspondence Analysis (CA) based pansharpening model which treats 6-band MS data ($LRM_2$, …, $LRM_7$) as a whole. The model is illustrated in Figure 3 and is accomplished by the following five steps:

Step 1: The generalized definition of the intensity component for n-band MS imagery can be usually represented by

$$I_{SYN} = \sum_{i=1}^{n} \omega_i LRM_i + b \tag{2}$$

where $I_{SYN}$ is the synthetic intensity of MS data. The set of coefficients $[\omega_2, \ldots, \omega_7]$ and $b$ can be directly calculated through multiple regression analysis of the original panchromatic image $PAN$ and the multispectral bands $LRM_i$ ($i = 2, \ldots, 7$):

$$PAN = \sum_{i=2}^{7} \omega_i LRM_i + b \tag{3}$$

Step 2: Assume that the spectral responses of the data set are practically unaffected by the change of the spatial resolution, the spatial details $\delta$ of PAN image can be extracted from $PAN$ and its low-passed version $P_L$.

$$\delta = PAN / P_L \tag{4}$$

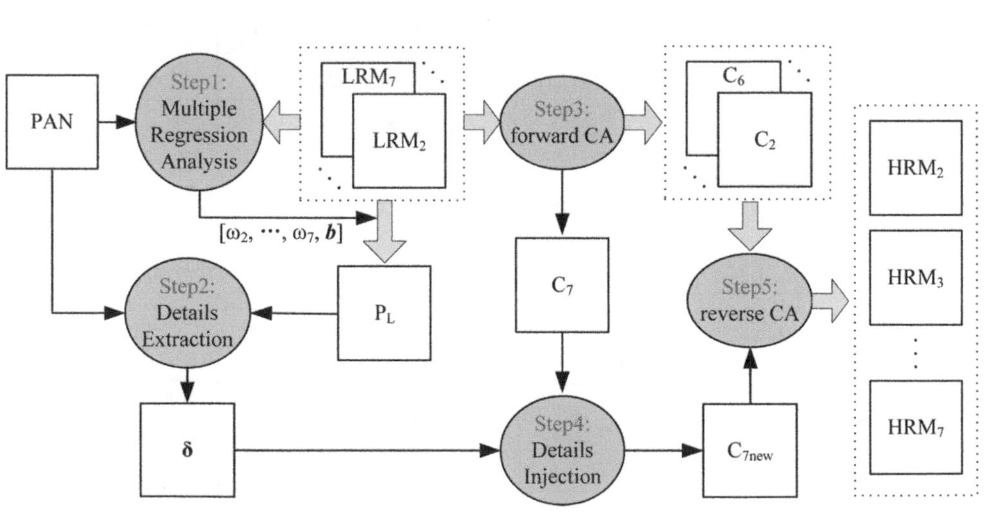


Figure 3. Schematic diagram of Model-2.

where $P_L$ is estimated from Equation 3.
Step 3: Transform the six band MS data ($LRM_2$, ..., $LRM_7$) into CA component space (forward CA transform) and obtain six component images ($C_2$, ..., $C_7$).

According to the CA approach (Cakir et al. 2006), the data table X is transformed into a table of contribution to the Person chi-square statistics. First, pixel $x_{ij}$ values are converted to proportions $p_{ij}$ by dividing each pixel $x_{ij}$ value by the sum $x_{++}$ of all the pixels in data set. The result is a new data set of proportions $Q$ and the size is row × column. Row weight $p_{i+}$ is equal to $x_{i+}/x_{++}$, where $x_{i+}$ is the sum of values in row $i$. Vector $[p_{i+}]$ is size of row. Column weight $p_{+j}$ is equal to $x_{+j}/x_{++}$, where $x_{+j}$ is the sum of values in column $j$. Vector $[p_{+j}]$ is size of column.

The Person chi-square statistic $\chi^2$ is a sum of squared $\chi_{ij}$ values, computed for every cell $ij$ of the contingency table.

$$\chi_{ij} = \sqrt{x_{++}}\left[\frac{p_{ij} - p_{i+}p_{+j}}{\sqrt{p_{i+}p_{+j}}}\right] \tag{5}$$

Use $q_{ij}$ values to form the matrix $Q_{row \times column}$, which is

$$Q_{row \times column} = [q_{ij}] = \left[\frac{p_{ij} - p_{i+}p_{+j}}{\sqrt{p_{i+}p_{+j}}}\right] \tag{6}$$

The matrix of eigenvector is given by

$$U_{column \times column} = Q^T_{column \times row} Q_{row \times column} \tag{7}$$

The 6-band MS data are transformed into the component space using the matrix of eigenvectors.
Step 4: Inject the details $\delta$ into the last component $C_7$.

$$C_{7new} = C_7 \times \delta \tag{8}$$

Step 5: Substitute the last component $C_7$ with the new one $C_{7new}$ and transform the new data set back into the original data space (reverse CA) to produce the sharpened MS images ($HRM_2$, ..., $HRM_7$).

## 3 EXPERIMENTAL RESULTS

### 3.1 *Data set and evaluation indices*

The testing WV-2 data set consists of 8-band multispectral imagery with 2.0 m spatial resolution and one panchromatic image with 0.5 m spatial resolution. The data were taken in Dec. 2009, which cover the urban area of Rome, Italy. The original size of the scene is 3200 × 3200 pixels, whose RGB composition (Band 5-3-2) is reported in Figure 4. One selected subscene of 800 × 800 pixels is shown in Figure 5(a), which constitutes vegetations, cars, buildings, roads, and shadow regions. The full-resolution PAN image of the subscene is reported in Figure 5(b).

Some commonly known global objective quality indices, including Correlation Coefficient (CC), Spatial Correlation Coefficient (sCC), Erreur Relative Globale Adimensionnelle de Synthèse (ERGAS), Spectral Information Divergence (SID), and Spatial Frequency (SF) are adopted to evaluate the pansharpened products (Chang 1999, Agrawal & Singhai 2010, Kalpoma et al. 2013, Marcello et al. 2013). For comparison purpose, APS, HCS, and GRSR-PDI (Zhang & Zhang 2011) methods are also implemented and tested.

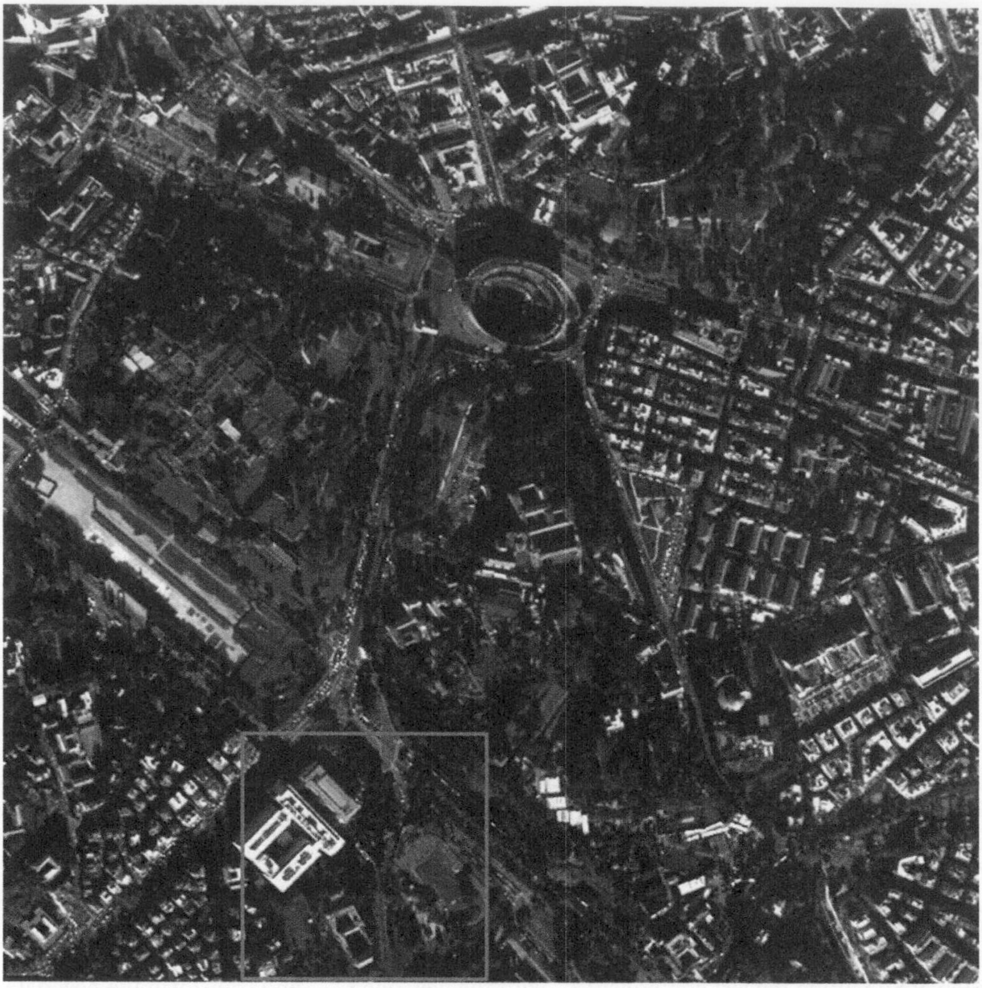

Figure 4. Original WV-2 MS natural color image (Band 5-3-2).

### 3.2 *Visual analysis*

The pansharpened results obtained from the proposed, APS, HCS, and GRSR-PDI methods are reported in Figures 5(c)–(f), respectively. By visually comparing the spatial quality of all the results with that of the MS image (Fig. 5(a)), it is obvious that the spatial resolutions of all the results are enhanced. Some small objects, such as cars, trees and roofs, which are not interpretable in the original image, can be indentified individually in each of the resultant images. Buildings and roads are much sharper in the resultant images than the original MS image. The most noticeable spectral distortions occur in Figure 5(f). The color hue of the green areas, especially on the grass within the playground and the roof of building, changes evidently. In Figure 5(e), the whole image appears blurred and some artifacts around the edges of the bright roof parts are introduced. It can be found that GRSR-PDI method produces color distortion obviously although it keeps spatial information well and HCS method injects spatial information insufficiently. The spatial improvement of APS method, shown in Figure 5(d), is impressive; however, slight color distortions exhibit on the vegetations and shadow parts with respect to the MS image of Figure 5(a). The result of the proposed method is shown in Figure 5(c). The spatial details appear nearly as sharp as those of PAN

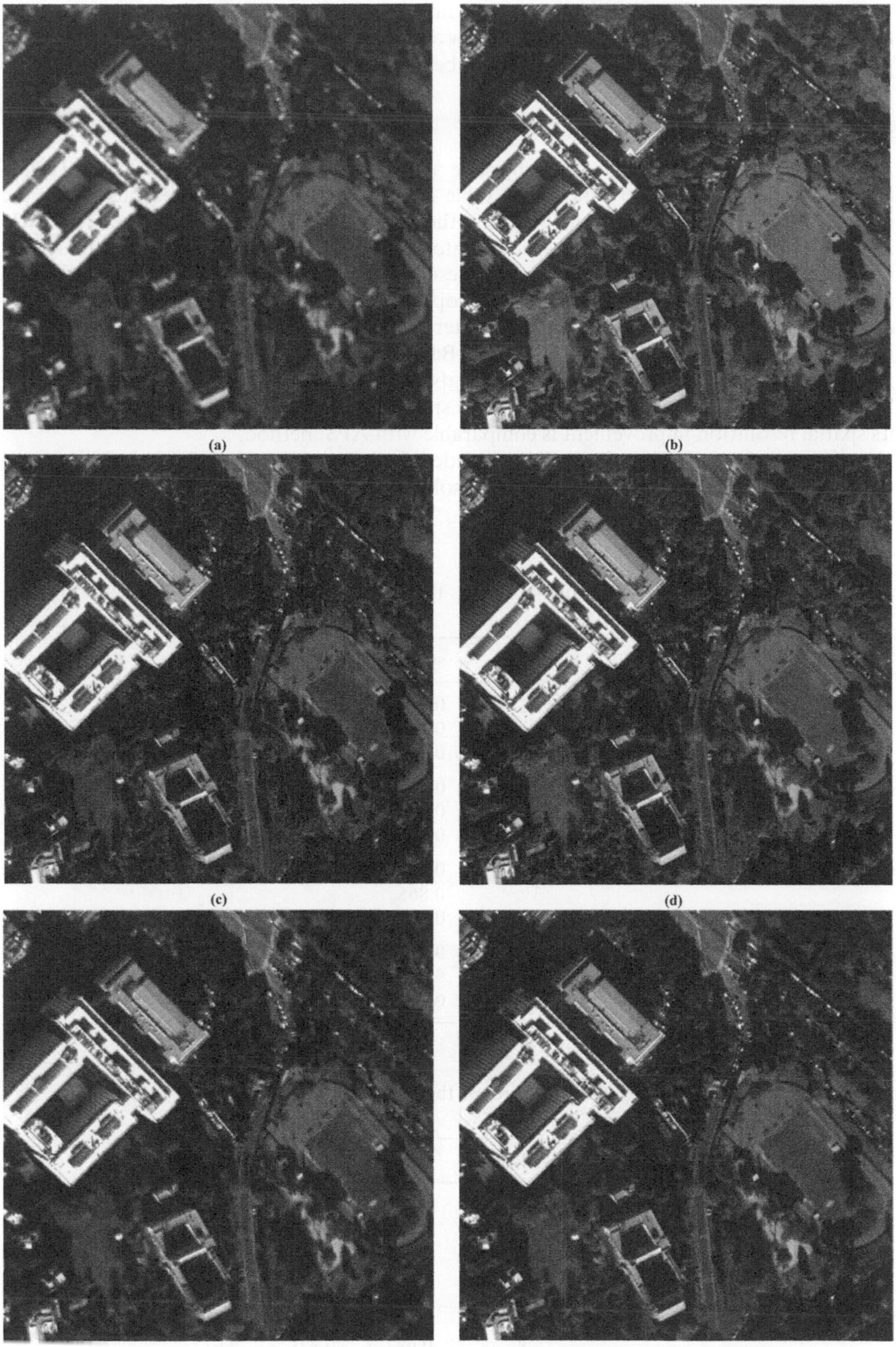

Figure 5. 800 × 800 subscene of (a) MS image, (b) PAN image, (c) the proposed method, (d) APS method, (e) HCS method, (f) GRSR-PDI method.

image (Fig. 5(b)) and spectral information is faithfully preserved without any obvious color distortion, such as the tree crowns and shadow areas. It can be seen that the proposed method produce high-quality image which shows the best visual effect on the whole.

### 3.3 *Objective analysis*

Visual judgment is corroborated by objective analysis. Score values are reported in Table 1 for the original 3200 × 3200 data set. In Table 1, HCS and GRSR-PDI methods obtain the poor quality scores compared with APS and the proposed methods. From the sCC and SID indices, HCS method keeps more spectral information than GRSR-PDI, while the spatial improvement of GRSR-PDI is better. APS results the best spatial enhancement, however it produces more color distortions than the proposed method. The proposed method is very close to APS from the sCC and provides better performance and outperforms best in terms of ERGAS and SID indices. Evaluations on Band1 and Band8 are also carried out. Table 2 reports the evaluation results for the two bands produced by Model-1. From Table 2, it can be seen that the proposed method keeps good spectral relationship with the original data and its spatial resolution improvement is comparable with APS method.

As a general observation, it can be concluded that the proposed method works well and obtains a goo trade-off between the spatial resolution enhancement and the spectral information preservation.

Table 1. Quality scores of the pansharpened composite results (Band 5-3-2).

| | Band | sCC | ERGAS | SID |
|---|---|---|---|---|
| The proposed | 2 | 0.977 | 2.16 | 0.302 |
| | 3 | 0.985 | | |
| | 5 | 0.983 | | |
| APS | 2 | 0.981 | 2.25 | 0.353 |
| | 3 | 0.989 | | |
| | 5 | 0.991 | | |
| HCS | 2 | 0.880 | 6.89 | 0.356 |
| | 3 | 0.865 | | |
| | 5 | 0.850 | | |
| GRSR-PDI | 2 | 0.940 | 4.32 | 0.520 |
| | 3 | 0.953 | | |
| | 5 | 0.950 | | |

Table 2. Quality scores of the pansharpened Band1 and Band8.

| | Band | CC | sCC | SF |
|---|---|---|---|---|
| The proposed | 1 | 0.954 | 0.958 | 17.1 |
| | 8 | 0.968 | 0.953 | 18.5 |
| APS | 1 | 0.932 | 0.967 | 17.5 |
| | 8 | 0.969 | 0.955 | 18.8 |
| HCS | 1 | 0.966 | 0.880 | 6.96 |
| | 8 | 0.982 | 0.825 | 9.81 |
| GRSR-PDI | 1 | 0.923 | 0.921 | 7.42 |
| | 8 | 0.952 | 0.885 | 7.86 |

## 4 CONCLUSIONS

A novel band grouping pansharpening method for WorldView-2 imagery is presented in this paper. Based on the spectral relationship between PAN and MS bands, the input MS images are grouped into two parts. Then two different pansharpening models are individually developed to process the grouped data. The experimental results show that the proposed method performs well on WV-2 data set that contains mixed areas. Visual analysis and objective evaluation demonstrate that the proposed method can achieve a better global fusion quality and is promising.

## REFERENCES


Agrawal, D. & Singhai, J. 2010. Multifocus image fusion using modified pulse coupled neural network for improved image quality. *IET Image Processing* 4(6): 443–451.

Aiazzi, B. et al. 2002. Context-driven fusion of high spatial and spectral resolution images based on oversampled multiresolution analysis. *IEEE Transactions on Geoscience and Remote Sensing* 40(10): 2300–2312.

Cakir, H.I. et al. 2006. Correspondence analysis for detecting land cover change. *Remote Sensing of Environment* 102: 306–317.

Camps-Valls, G. et al. 2011. Introduction to the issue on advances in remote sensing image processing. *IEEE Journal of Selected Topics in Signal Processing* 5(3): 365–369.

Chang, C.I. 1999. Spectral information divergence for hyperspectral image analysis. *IEEE Proceedings of IGARSS* 1: 509–511.

Choi, M. et al. 2005. Fusion of multispectral and panchromatic satellite images using the curvelet transform. *IEEE Geoscience and Remote Sensing Letters* 2(2): 136–140.

Ji, M.H. et al. 2012. Improving spectral fidelity of WorldView-2 image fusion via a constrained generalized intensity-hue-saturation model with localized weight structure through land cover classification. *Journal of Applied Remote Sensing* 6(1), 061707: doi:10.1117/1.JRS.6.061707.

Kalpoma, K.A. et al. 2013. IKONOS image fusion process using steepest descent method with bi-linear interpolation. *International Journal of Remote Sensing* 34(2): 505–518.

Laben, C.A. & Brower, B.V. 2000. Process for enhancing the spatial resolution of multispectral imagery using pansharpening. *USA Patent* 6011875.

Liu, J.G. 2000. Smoothing filter-based intensity modulation: A spectral preserve image fusion technique for improving spatial details. *International Journal of Remote Sensing* 21(8): 3461–3472.

Longbotham, N. et al. 2012. Very high resolution multiangle urban classification analysis. *IEEE Transactions on Geoscience and Remote Sensing* 50(4): 1155–1170.

Mahyari, A.G. & Yazdi, M. 2011. Panchromatic and multispectral image fusion based on maximization of both spectral and spatial similarities. *IEEE Transactions on Geoscience and Remote Sensing* 49(6): 1976–1985.

Marcello, J. et al. 2013. Evaluation of spatial and spectral effectiveness of pixel-level fusion techniques. IEEE *Geoscience and Remote Sensing Letters* 10(3): 432–436.

Padwick, C. et al. 2010. WorldView-2 pan-sharpening. *ASPRS 2010*, San Diego, California.

Tu, T.M. et al. 2001. A new look at IHS-like image fusion methods. *Information Fusion* 2(3): 177–186.

Tu, T.M. et al. 2012. An adjustable pan-sharpening approach for IKONOS/QuickBird/GeoEye-1/WorldView-2 imagery. *IEEE Journal of Selected Topics in Applied Earth Observations and Remote Sensing* 5(1): 125–134.

Wang, Z. et al. 2005. A comparative analysis of image fusion methods. *IEEE Transactions on Geoscience and Remote Sensing* 43(6): 1391–1402.

Zhang, D.M. & Zhang, X.D. 2011. Pansharpening through proportional detail injection on generalized relative spectral response. *IEEE Geoscience and Remote Sensing Letters* 8(5): 978–982.

Zhang, Y. 1999. A new merging method and its spectral and spatial effects. *International Journal of Remote Sensing* 20(10): 2003–2014.

*Information Systems and Computing Technology – Zhang & Gu (eds)*
*© 2013 Taylor & Francis Group, London, ISBN 978-1-138-00115-2*

# Research on GIS based haze trajectory data analysis system

Yuanfei Wang
*Key Laboratory of Geo-information Science, Ministry of Education, East China Normal University, Shanghai, P.R. China*

Jun Chen
*Taobao (China) Software Co., Ltd., Hongzhou, Zhejiang, P.R. China*

Jiong Shu
*Key Laboratory of Geo-information Science, Ministry of Education, East China Normal University, Shanghai, P.R. China*

Xingheng Wang
*Computing Center of East China Normal University, Shanghai, P.R. China*

ABSTRACT: Recent years, haze, an atmospheric phenomenon, has developed into a primary enemy of urban air pollution that has been given great importance by many cities in China. In consideration of the complication of process of haze, the inquiry system for haze trajectory based on Web and GIS, is established for realizing an integrated application circumstance of Multi-sourced heterogeneous data, this includes data from haze trajectory, meteorological condition, environmental quality, backscatter, weather image, satellite cloud imagery etc. In this platform, we not only skim through and analyze the data of single factor, but also realize conjoint analysis of multi-dimension data, which is a useful trail to studying the process of haze, obtaining relevant knowledge, sharing relevant data, and to forecasting the arrival of haze.

## 1 INTRODUCTION

Haze is an atmospheric phenomenon where dust, smoke, and other dry particles obscure the clarity of the sky, resulting in horizontal visibility of less than ten kilometers [China Meteorological Administration, 2003]. The main difference of haze and fog lies in the water vapor content, generally it is regarded as necessary of haze when the relative humidity is 90% [Dui Wu, 2008]. With the rapid development of China's economy, obvious rising of energy consumption, sharp increase of the number of automobile ownership, and the outward expansion of urban construction, great changes have taken place in the situation of air pollution, this is exemplified by the pollution of photochemistry and haze with total suspended particle is becoming increasingly severe. The haze weather pollutant composition may also have some difference in different areas. For example, the main pollutants of Xiamen is $SO_2$, $NO_x$, PM10 [Xinqiang Fan, Sunzhao Bo. 2008], and the high concentration of pollutants in Pearl River delta including CO additional [Xunlai Chen, Yerong Feng, etc 2007]. The polluted area which ozone and haze cover has formed in Beijing-Tianjin-Hebei Province Region, Yangtze River Delta, Pearl River Delta and Chengdu-Chongqing Region, and the haze has become a new meteorological and environmental disaster [Xiaoqing Rao, Feng Li. 2007] [Yue Wu, 2007]. The 2011 Annual Environmental Status Bulletin of Shanghai was shown that the problem of chemical and haze pollution has become an important factor which restricts environmental quality of Shanghai. Therefore, the research and forecast for haze pollution has become the hotspot of protection of urban air quality in recent years [Limei Jin, Jun Shi, 2008] [Demin Shao, Wei Zhang, etc 1992].

Haze studies have been carried out by foreign and domestic experts in terms of the aspect of its formation mechanism, meteorological condition and process, climate features and spatial and temporal distribution, many of them study in chemical analysis. For example, Min Min on ash haze process characteristics of aerosol [Min Min, Pucai Wang, etc 2009], Jue Ding, Liying Liu to fog haze weather particle contaminant characteristics research [Jue Ding, Liying Liu 2009], etc. It is a new trend to study the analysis of its source and the process of its pollution through remote sensing, laser, new observation technique, as well as air mass trajectory model [Yan Wang, Fahe Chai etc, 2008]. Jiong Shu uses air mass trajectory model contributing to source-tracing and simulation of haze pollution demonstrate that haze emerging in city can be caused, in favorable meteorological condition, by the interaction force between local air mass and external air mass [Yong ping Long, Jiong Shu, 2010]. According to this living example of Shanghai, the external including the haze pollution mainly comes from Yangtze River Delta, which objectively shows that haze pollution is a kind of weather phenomenon that regionally emerges in metropolitan areas. When haze pollution arises in a city, the visibility in the region that the whole metropolitans covers may remain low. But this deduction will need to be further probed.

That Haze is not favorable to urban traffic safety and body health and is one of the serious environmental issues has been demonstrated by many researches, within the definition of subject to haze still have certain differences, especially in the relative humidity [Guolian Liao, Peng Zeng, etc, 2008] [WMO, 1996] [Fang Wang, DongSheng Chen, Shuiyuan Cheng etc, 2009]. Although numerous researches are being tried to establish early warning system, few reliable system is established due to the complicated action mechanism of haze pollution. An inquiry system, which has been established, is essential an integrated data analyzing for haze pollution. And it works by applying technique of database to acquire meteorological data of urban meteorological observation, satellite cloud imagery and weather image data that serve for the formation of air pollution. Afterwards, by managing and analyzing backward trajectories data for haze pollution based on HYSPLIT model computation and air quality data of urban meteorological observation and also by applying electronic maps, realize the visibility of Web and GIS application of haze pollution trajectory, and of analysis of relevant integrated data.

## 2 SYSTEM DATA

As previously mentioned, inquire and analysis system for haze trajectory centralized by locus of motion of polluted air mass of haze integrates meteorological observation data with satellite cloud imagery and environmental monitoring data in order to provide data service for relevant studies, data analyzing service for forecast, and also the knowledge dissemination service for the citizens. Besides, with furnishing it with real-time analyzing online function of HYSPLIT model access to Web, this system has certain degree functionality of forecast.

The data is the basis of the system. The purpose of the system as well as the accessibility of data should be taken into consideration when data composition of system is established. This system integrates various data, including haze trajectory data, meteorological and air quality observation data, weather image data, satellite cloud data, backscatter data, stratification curve.

### 2.1 *Haze trajectory data*

haze trajectory data, obtained by calculation of HYSPLIT, describes the source of air mass caused by haze pollution in a region or a city. The certain number of haze, and the information of latitude and longitude specific time are included in this database, which are visualized on the screen of electronic map, showing motion condition of polluted air mass of haze.

### 2.2 *Meteorological and air quality data*

A certain number of meteorological conditions need to be involved with the emergence and differentiation of meteorological phenomenon of haze, and physical constitution and urban environmental quality are closely related, which is the essential basis for establishment of system. Contributing factors of haze pollution in Shanghai can be deducted by that system alternatively integrates meteorological data of shanghai with data of air quality observation, combined with haze trajectory data. Meteorological data includes wind velocity, humidity, temperature, visibility etc. while the data volume of air quality is far more than the meteorological data, including PM10, PM2.5, API, Pb, Sr, In, Cn, Ni, Fe, Mn, Cr, Ti, S, Al, Mg, Na, F-ion, Cl-ion, NO2-ion, SO4-ion, Na-ion, NH4-ion, K-ion, Mg-ion, Ca-ion, etc.

### 2.3 *Weather image*

Ti is indispensable data in analyzing weather condition during the course of haze pollution. The process of emergence, development, motion of weather system is analyzed with different heights through weather image, which has main effect to formation of haze, and distribution and change of pollutants generated in the course of haze pollution. Six kinds of weather images in different heights, stored in data, which are composed of ground, 920 hPa, 850 hPa, 700 hPa, 500 hPa, 300 hPa, and sampling time varies from heights to heights. Starting at zero hour of per day, the weather image of ground is recorded at intervals of 3 hours, and the time for recording. However, weather images of remaining heights respectively are zero hour and 12 hours in one day.

### 2.4 *Satellite cloud imagery*

The imagery of cloud covering and surface features can be observed by meteorological satellites that can identify the emergence, development and evolution of weather system, combined with haze trajectory data contributing to the forecast and analysis of haze.

### 2.5 *Backscatter ratio imagery*

It obtained by the measurement of ceilometer CL31, is a way of calculating visibility. And the instrument is placed in the site of 31.13°N, 121.23°E for continuous observation of 24 hours.

### 2.6 *Stratification curve imagery*

It shows the vertical change of meteorological elements, by using this data, wind direction, wind velocity, relative humidity of low earth atmospheric layer and atmosphere structure of boundary layer, and these are important evidence for judging emergence of haze pollution, as well as the evidence for causing haze pollution. The system records stratification curve data at zero hour and 12 hour in Shanghai in one day.

## 3 SYSTEM ARCHITECTURE DESIGN

The inquiry and analysis system for haze trajectory is based on application system of Web, and is framed in a way of B/S (browser/server). The browser is responsible for access to system and operating system, and the server is in charge of undertaking complicated calculations and inquiry service and returning results. The function of information's publication of Internet and the feature of data sharing are greatly exerted. With this design, the users can be available to the system without installing any software and database. In this way, a good platform is constructed for integrated data analysis and data sharing service. Figure 1 is the framework of the system.

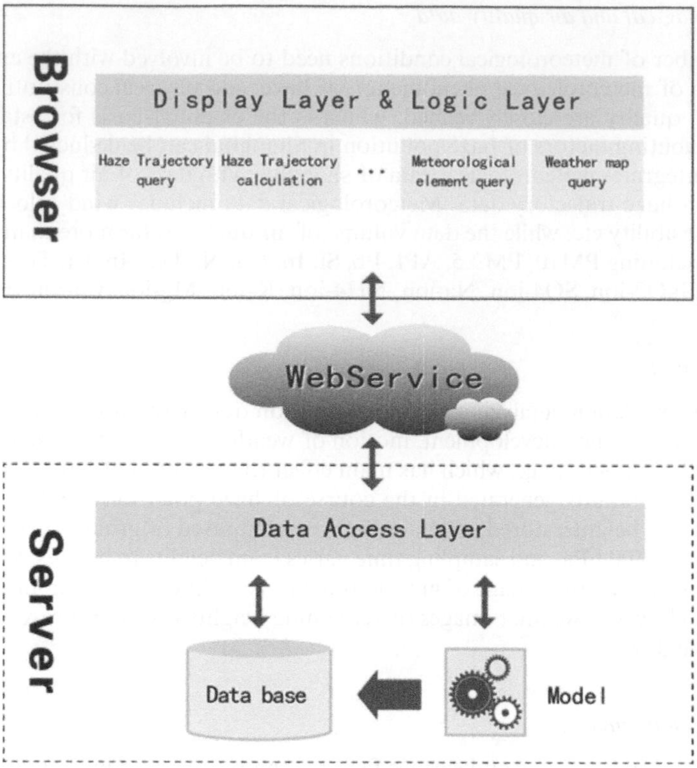


Figure 1. The framework of the system.

According to the frame structure of the system seen in Figure 1, the system is composed of client layer, business logic layer, Web service, data access layer and data layer.

### 3.1 *Clients*

As the bridge of interaction between the users and the system, namely users' request to business logic layer and receives feedback from business logic layer to show the users results.

### 3.2 *Business logic layer*

This refers to that according to different requests, it can use related function modules to respond to the request from client layer. As for the requests that are related data operating order, the system can carry out access to server through web service interface.

### 3.3 *Web service*

By encapsulating the function provided by server for the user's use, and transporting the requests from clients and the results of server through HTTP, the communication between clients and server can be realized.

### 3.4 *Access data layer*

Data storage is abstractly encapsulated to realize the effective use, processing and analysis of underlying data source and to answer to the requests of system's data service.

### 3.5 *Data layer*

As the data source of system, data layer, located in the bottom of system of system, is established to answer the requests from access data and to maintain the integrity and safety of system.

## 4 TECHNOLOGY OF SYSTEM IMPLEMENT

Development environment: the system based on the development environment of Microsoft Visual Studio takes C# as the main development language in background and integrates with Flex technique, which enhances the interaction and aesthetics of system. As an efficient and free open source framework, its higher column picture can execute the highest data-intensive application, attaining to the execution speed of desktop application program. Besides, it is used for constructing the web application program with performance and user experience map search engine. The system employs Google Earth as map service. Google Map is free electronic map service provided by Google, which provides global world with map researching and zoom function. With fast loading rate, Google Map encapsulates some basic operation of the map which is convenient for secondary development of GIS system. It is convenient to use Google Map API and Dom for loading the information of haze trajectory on the basis of Google Map serving for map service.

Microsoft SQL Server 2005 is applied to database. The DBMS software of system, which is favorable to efficient access to data and ensures the safety of data with the function of log file records for database operating and the function of backup and restore.

According to the data type related to the data of system, data can be divided into two kinds of ways of storing, among which haze trajectory data and observation data are stored in database as bivariate table. The format picture data like weather image and satellite cloud imagery are stored in the specified files, and only the filename and the time and other information are stored in database tables. The system is composed of trajectory information, image information, meteorological data, pollution data, and user information. The relationship is shown in Figure 2.

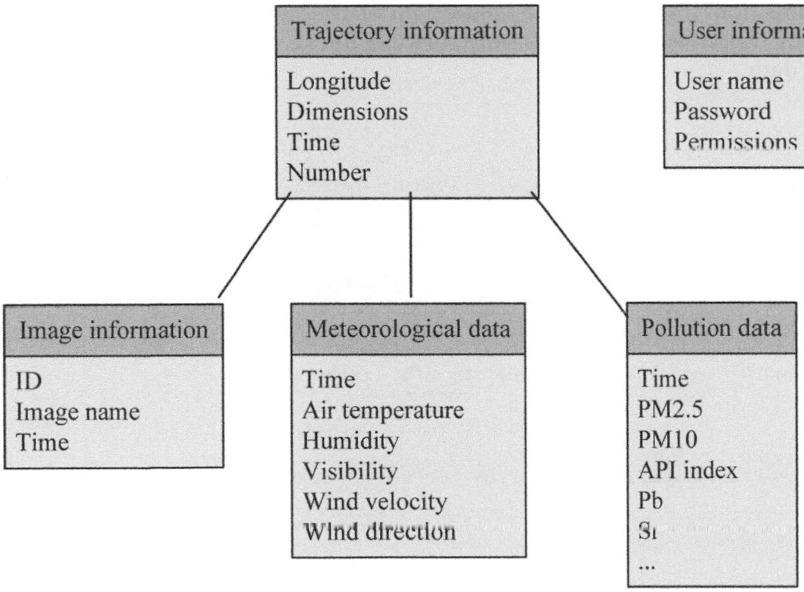


Figure 2. Table structure of database.

HYSPLIT, applies to the calculation of air mass trajectory, initially was jointly developed by ARL of NOAAC (US national Oceanic and Atmospheric Administration) and Australia's Bureau of Meteorology. The version 4.9 used in this system described in this dissertation, was updated in January 2010. This model adopts the approach that mingles Euler's with Lagrangian's, and the former approach was applied to concentration calculation, the latter to the calculation of advection and diffusion [Roland R. Draxler, 1999]. Suppose the motion path of particle depends on wind field, so the location vector of particle trajectory is the integration between time and space, and speed vector is bilinear interpolation in time and space. Suppose the speed of particle is $V$ which is on original position $P$, then after time steps of $\Delta t$, $P'$ is the primary estimated position.

$$P'(t+\Delta t)=P(t)+V(P,t)\Delta t$$

After time steps of $\Delta t$, $P\,(t+\Delta t)$, the ultimate position of particle can be calculated by the average value between its original position $P$ and three-dimensional velocity rector of $P'$.

$$P(t+\Delta t)=P(t)+0.5\left[V(P,t)+V(P',t+\Delta t)\right]\Delta t$$

That motion distance in each time step is less than grid resolution is prerequisite for the change of time steps of integration. The data on integration position is calculated by linear interpolation, and higher integration can not improve calculating precision. As a result, the course of calculation for ultimate position of particle only uses the primary estimated position.

The haze trajectory data can be acquired through two methods. One is that backward trajectories of haze pollution in heights of 500 m, 1500 m, 300 m (respectively correspond to 850 hPa, 700 hPa and 500 hPa), was calculated by HYSPLIT model. The other is that the user calculates online. It should be noted that HYSPLIT is a desktop model. For the realization of real-time calculation of Web-client, it is needed to make HYSPLIT as a kermel, with a layer wrapping up outside that you can enter with data, driving HYSPLIT through access to Web service. The trajectory data obtained by calculating automatically in database was read by Web access to database. The parameters, including location, time, trajectory direction, trajectory length, and height of trajectory, should be set for real-time calculating of trajectory. Besides, the trajectory can not be calculated without the support of air mass data.

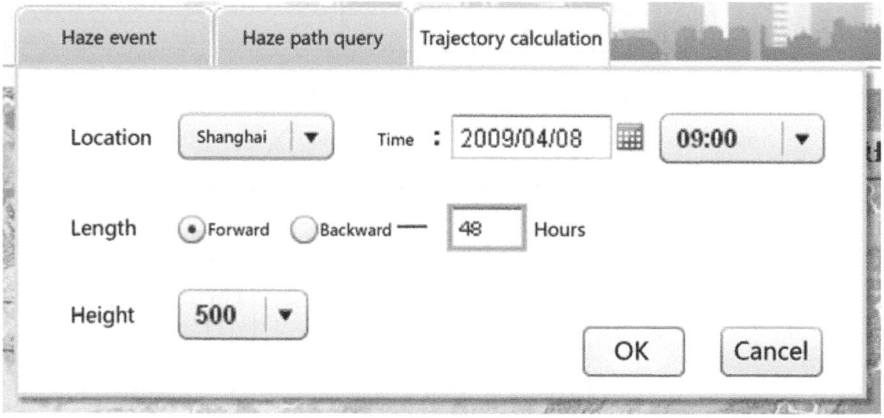


Figure 3. Two ways to acquire date of backward trajectories.

## 6 THE TECHNOLOGY AND APPLICATION OF CONJOINT ANALYSIS OF MULTI-DIMENSION DATA

The study shows that haze pollution is a complicated weather, which is related to meteorology, urban environmental pollution, even economy of the metropolitans and other interacted factors. Single data, incapable of explaining the cause of haze, can not forecast the formation of haze development. Under the condition of sufficient data, data excavation is a kind of analysis technique in accordance with complicated phenomenon and cause. At the beginning of this research, the conjoint analysis of multi-dimension data was considered as a main technique for disclosing haze pollution, which is reflected by interface layout in Figure 4. In the design function, haze trajectory, meteorological and environmental, weather map, satellite cloud imagery, and backscatter, can be applied to the conjoint analysis like the companion and overlapping of data. This conjoint analysis of multi-dimension is demonstrated by two living examples blow.

### 6.1 *Practical example 1: The conjoint analysis of meteorological air quality data in the course of haze pollution*

For each haze trajectory, the information will be uploaded to monitoring site around it, and the geographical and air quality data is convenient for spotting the relationship of spatial and temporal distribution among factors. The user can choose to examine the hourly data or daily average data.

### 6.2 *Practical example 2: Related analysis of dynamic process of haze trajectory and satellite cloud imagery*

When the user need to look over satellite cloud imagery, the system will automatically retrieve all imagery in the specific time of haze pollution, and put in on the Google electronic map,

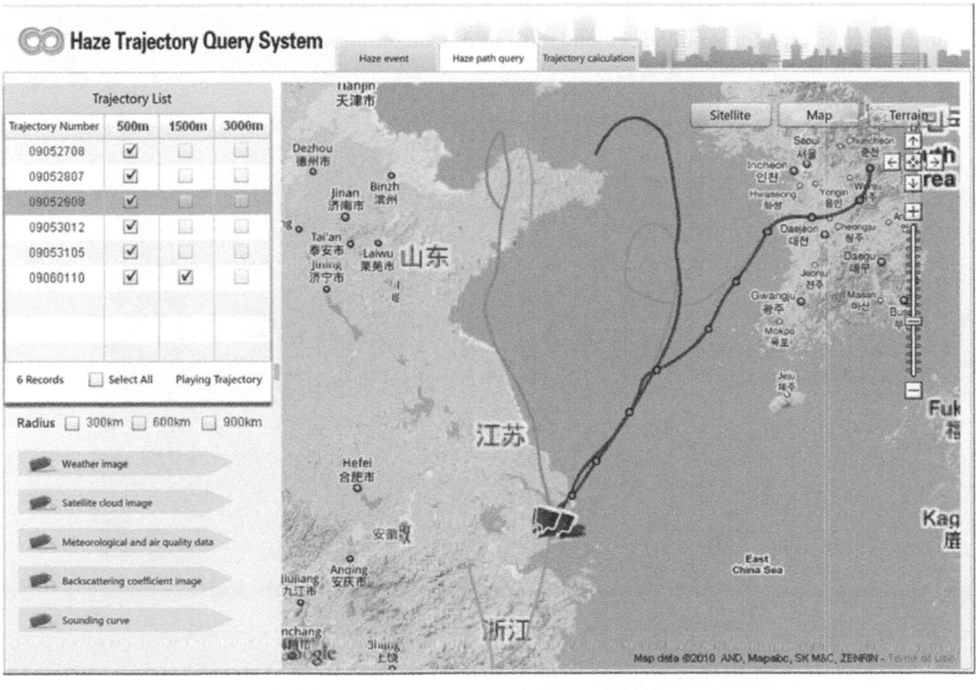


Figure 4. User interface of the system.

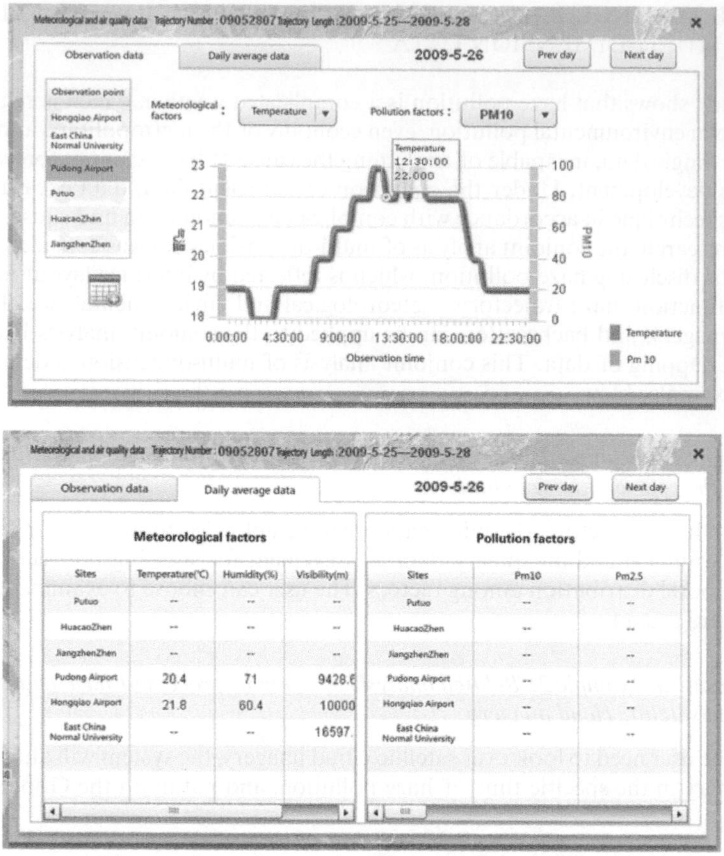


Figure 5. Haze trajectory and the conjoint analysis of meteorological air quality data.

Figure 6. Overlay analysis of haze trajectory and satellite cloud imagery.

forming intuitive weather map. The user can watch development course of satellite cloud imagery with real-time image, as well as certain transient imagery. Motion trajectory of haze air mass can be overlapped on satellite cloud imagery in chronological sequence of satellite cloud imagery haze trajectory. The information of weather in large scope in the course of haze pollution can be acquired by the user through this image addition technique.

## 7 CONCLUSION

This research establishes a inquire system for haze trajectory on the basis of Web & GIS and other techniques, which realize the real-time online of backward trajectories, the calculating model of Web, and the goal of conjoint analysis and application of multi-dimension data, and provides a comprehensive data integrating environment for the study of urban haze pollution, data sharing service, learning haze pollution knowledge. The result judgment based on data analysis also conforms to some research [Dui Wu, 2008] [Jun Shi, Wei Wu, 2010] [Fang Wang, Dongsheng Cheng, 1999] [Biyun Zhao, Bin He, etc, 1999]. As a kind of information technology applied in a new field, this system has a large space for development, such as the integrity of haze pollution with data type and more data of observation, the enhancement of data in statistics and analysis, and other data analysis and model integration [Dai J.M, Rocks D.M, 2000] [R. Sivacoumar, A.D. Bhanarkar, 2001] [H. Björnsson, S.A. Venegas, 1997]. What have mentioned above is also the future development direction of this platform.

## REFERENCES


Biyun Zhao, Bin He, Faqing Zhu. Atmospheric pollution dispersion spatial information systems [J]. Environmental Sciences, 1999, 12(6):10–12.

Björnsson H., Venegas S.A. A manual for EOF and SVD analyses of climatic data. [E]. http://www.geog.mcgill.ca/gec3/wp-content/uploads/2009/03/Report-no.-1997-1.pdf.1997.2.

China Meteorological Administration. Ground meteorological observation specification [S]. Beijing: China meteorological press: 2003, 21–27.

Demin Shao, Wei zhang, Aihua Shen. Shanghai upper airflow long-distance transport of trajectory and the relationship between the acid rain [J]. Shanghai: Shanghai environmental science. 1992, 11(1):6–9.

Dai J.M, Rocks D.M.A GIS2 based approach to spatial allocation of area source solvent emissions. Environmental Modeling & Software, 2000, 15:293–302.

Dui Wu. Haze and fog identification and data analysis process [J]. Journal of environmental chemistry. 2008, 27(3): 327–330.

Guolian Liao, Peng Zeng, Peng Cheng. Pearl river delta typical grey haze weather process and clean too close to laminar flow field of a comparative analysis of the EOF [J]. Weather Research and Application. 2008, 29(4): 23 to 25.

Fang Wang, Dong sheng Cheng, Shuiyuan Chen, Mingjun Li. Transportation affect the airflow trajectory clustering-based air pollution [J]. Environmental Sciences, 2009, (06).

Fang Wang, Dong Sheng Chen, Shuiyuan Cheng etc. Atmospheric pollution transfer influence Based on airflow trajectory clustering [J]. Journal of environmental science research. 2009, (6): 637–642.

Jue Ding, Liying Liu. Fog haze weather particle pollutant characteristics and absorption of gaseous contaminants process analysis [J]. Shanghai: Shanghai environmental science. 2009, 28(4):11–14.

Jun Shi, Wei Wu. Haze serial reconstruction and its spatial and temporal characteristics of the climatic data of Shanghai [J]. Resources and Environment of the Yangtze River Basin, 2010, (09).

Limei Jin, Jun Shi. Fog and haze day's climate characteristics and change rule of Shanghai [J]. J plateau weather. 2008, 27 (supplement): 138–143.

Min Min, Pucai Wang, Xuemei Zong etc. Gray haze in the process of the aerosol characteristic observation study [J]. J climate and environmental research. 2009, (2):153–160.

Roland R. Draxler. Description of the HYSPLIT_4 Modeling System//NOAA Technical memorandum ERL ARL-224.1999.

Sivacoumar R., Bhanarkar, A.D. Air pollution modeling for an industrial complex and model performance evaluation. Environmental Pollution, 2001, 111:471–477.

WMO. Guide to Meteorological Instruments and Methods of Observation [s]. 1996 (6th medition). 1.14–3.

Xiaoqing Rao, Feng Li. Analysis of a big range of Haze [A]. Chinese Meteorological Society 2007 Annual Meeting of atmospheric composition observations, research and forecasting sub-venue, 2007.
Xinqiang Fan, Sunzhao Bo. 1953–2008 Xiamen region gray haze weather characteristics [J]. Journal of atmospheric sciences. 2009, (5):604–609.
Xunlai Chen, Yerong Feng, An Yu Wang, etc. Numerical study of ash haze weather major pollutants in the Pearl River delta urban agglomeration [J]. Journal of sun yat-sen university. 2007, 46–48 (4): 103–107.
Yan Wang, Fahe Chai, Houfeng Liu etc. Atmospheric pollutant level transport field characteristic analysis [J]. Journal of environmental science research of the Yangtze river delta. 2008, (1):22–29.
Yongping Long, Jiong Shu. The Backward Trajectories Analysis of Typical Haze in Shanghai [A]. The seventh Yangtze River Delta meteorology technology forum, 2010.
Yue Wu. Metropolitan areas the difference between haze and fog and haze weather warning [A]. Chinese Meteorological Society, 2007. Annual Meeting of atmospheric composition observations, research and forecasting sub-venue, 2007.

*Regular papers*

*Information Systems and Computing Technology – Zhang & Gu (eds)*
*© 2013 Taylor & Francis Group, London, ISBN 978-1-138-00115-2*

# A warning model of systemic financial risks

Wenbin Xu
*School of Economics and Management, Beijing Information Science and Technology University, Beijing, China*

Qingda Wang
*Department of International Economics and Trade, University of International Relations, Beijing, China*

ABSTRACT: In recent years, the international financial situation is grim and financial disasters happen often and bring huge economic losses to the countries. The systemic financial risk is a main factor to undermine the financial stability so it forces the government especially the financial regulatory authorities to attach importance to. In this paper we build a warning model of the systemic financial risk based on fuzzy pattern recognition. With the model we predict the risk through a number of data integration and that is of great significance for the macro-control.

## 1 INTRODUCTION

In the last three decades international financial turbulence occurs frequently and brings huge economic losses to the countries. And it makes people become increasingly aware of that the financial security is the core of a country's economic security. To maintain financial stability as an important economic policy is a consensus among all countries. The systemic financial risk is a main factor to undermine the financial stability so it forces the government especially the financial regulatory authorities to attach importance to. Especially since the outbreak of the subprime mortgage crisis in the United States in 2007, financial systemic risk has once again aroused the lively discussion of governments, international organizations and academia. Since China's accession to the WTO, the pace of economic and financial reform and opening up continue to accelerate. The financial system is in a period of drastic changes and at the same time systemic financial risk also continues to accumulate. To predict the systemic financial risk is of increasingly importance for the prevention of financial crises. The difficulty of managing systemic financial risk lies in its correct and timely measure. Therefore, how to select the appropriate method to monitor changes in the systemic financial risk in order to prevent the occurrence of a systemic crisis is a very interesting research topic.

## 2 RESEARCH STATUS

In view of the continuous development of the external environment and business strategy, enterprise risk warning has become the focus of academia and business, so many experts or scholars at home and abroad have conducted in-depth study. Beginning in 1970, there exits some theories like strategic risk management and VaR assets assessment. Many foreign scholars analysis the early warning from the enterprise level and draw many useful conclusions. But the research is primarily aimed at the corporate crisis and how to deal with it. Cause of the crisis or the development process is the lack of analysis. Taken together, the early warning models are mainly the following ways.

- Multivariate statistical analysis;
- Artificial Neural Network Model;

- System dynamics method;
- Support Vector Machine Model.

## 3 EARLY WARNING MODEL BASED ON FUZZY PATTERN RECOGNITION

Fuzzy pattern recognition has a good application in the identification and early warning of systemic financial risk. The judgment of the systemic financial risk is the basis of macro-control. Depending on the level of risk we can take the appropriate policy. The complex and volatile financial markets pose a serious challenge to the detection of systemic financial risk. It increases the difficulty of detection by a single method so we predict the risk through a number of data integration.

### 3.1 *Feature sets*

There are many factors that affect the systemic financial risks and the features of each factor differ. From a general point of view, the features of the systemic financial risks can be summarized as follows.

$$X = \{B, C, L, E, P, IE, PP\}$$

where
$B$: The level of the banking industry for bad and doubtful debts
$C$: Rate of credit
$L$: Liquidity
$E$: Real estate financial bubble
$P$: Policy
$IE$: Interest rate & exchange rate
$PP$: Purchasing power.

### 3.2 *Conclusion sets*

The conclusion set is a collection of all the results may be obtained by the fuzzy pattern recognition, represented by $Y$.

$$Y = \{A_1, A_2, A_3, A_4, A_5\}$$

where
$A_1$: Extremely dangerous
$A_2$: Very dangerous
$A_3$: Dangerous
$A_4$: General
$A_5$: Non-dangerous.

### 3.3 *Standard form*

The feature sets $X$ is the problem domain and the elements in the conclusion sets $Y = \{A_1, A_2, A_3, A_4, A_5\}$ are the standard forms. Each conclusion set should be a collection of fuzzy sets on the feature sets $X$.

To establish the standard forms is to determine the membership function of $A_1$–$A_5$.

Due to the great quantity of influencing factors and complex relationships, how to determine the fuzzy membership functions has a lot of choices like expert assessment method, priori formula method and expert scoring method. In this paper, based on the expert scoring method we give the corresponding fuzzy membership function.

We assume that the five standard form membership functions determined by the expert scoring method are shown in the Table 1 (In this paper, the data were normalized).

### 3.4 *Fuzzy pattern recognition algorithm*

In this paper, we choose to use the weighted closeness degree, and then do the pattern recognition in accordance with the fuzzy selecting rule.

The first step is to identify a kind of closeness degree.

$$q(A,B)=\frac{1}{2}\{Hgt(A\cap B)+[1-Dpn(A\cup B)]\}$$

where

$Hgt\ A = sup\ \{A(x)|x\in X\}$ is the fuzzy set peak,

$Dpn\ A = inf\ \{A(x)|x\in X\}$ is the fuzzy set valley.

The second step, to determine the formula for calculating the measure of fuzzy set closeness degree.

After determined the weight, we calculate the closeness.

The weighted closeness degree is shown as

$$q(A^*,A_i)=\overset{7}{\underset{j=1}{X}} w_j q(A^*,A_{ij})$$

$j$, $A_{ij}$ are the i-th component of the $A^*$, $A_i$ where $A^*$ respectively.

The third step is to determine the weight.

It is a key work to determine the weight and the common ways are subjective method (expert scoring method) and object method (statistical experimental method).

In this case, we choose the weight vector to be

$$w=(0.09, 0.2, 0.12, 0.21, 0.1, 0.2, 0.08)$$

Table 1. Fuzzy pattern of the systemic financial risk level.

| | Feature | | | | | | |
|---|---|---|---|---|---|---|---|
| Form | B | C | L | E | P | IE | PP |
| $A_1$ | 0.40 | 0.50 | 0.45 | 0.50 | 0.40 | 0.55 | 0.25 |
| $A_2$ | 0.22 | 0.20 | 0.25 | 0.30 | 0.20 | 0.15 | 0.20 |
| $A_3$ | 0.18 | 0.12 | 0.10 | 0.15 | 0.20 | 0.15 | 0.20 |
| $A_4$ | 0.10 | 0.10 | 0.10 | 0.05 | 0.10 | 0.10 | 0.20 |
| $A_5$ | 0.10 | 0.08 | 0.10 | 0.00 | 0.10 | 0.05 | 0.15 |

Table 2. Result of the recognition.

| | Feature | | | | | | | |
|---|---|---|---|---|---|---|---|---|
| Form | B | C | L | E | P | IE | PP | $q(A^*, A_j)$ |
| $A_1$ | 0.036 | 0.09 | 0.0558 | 0.0945 | 0.045 | 0.085 | 0.0388 | 0.4451 |
| $A_2$ | 0.0441 | 0.08 | 0.0522 | 0.0945 | 0.045 | 0.075 | 0.0392 | 0.43 |
| $A_3$ | 0.0441 | 0.072 | 0.0432 | 0.07875 | 0.045 | 0.075 | 0.0392 | 0.3973 |
| $A_4$ | 0.0405 | 0.072 | 0.0432 | 0.06825 | 0.04 | 0.07 | 0.0392 | 0.3712 |
| $A_5$ | 0.0405 | 0.068 | 0.0432 | 0.063 | 0.04 | 0.065 | 0.0372 | 0.3569 |
| $W_i$ | 0.09 | 0.2 | 0.12 | 0.21 | 0.10 | 0.2 | | |

The last step is recognition. We extract the features of the influencing factors.

$$A^* = (0.2, 0.4, 0.38, 0.4, 0.3, 0.4, 0.22)$$

Then we can do the fuzzy pattern recognition. The result is shown in Table 2.

From the result table, we can identify that the systemic financial risk is extremely dangerous; we have to take appropriate measures to deal with it.

## 4 CONCLUSION

In this paper, we use feature extraction recognition algorithm, taking full use of the advantage of the fuzzy pattern.

Recognition method has high tolerance of the outside interference. We can take advantage of the experience and knowledge of experts, inducing it into inference rules and degree between the unknown object and the standard form and then do the pattern recognition in accordance with the fuzzy selecting rule, completing the recognition with it.

## ACKNOWLEDGMENT

Here and now, I would like to extend my sincere thanks to Beijing Planning Fund of Philosophy and Social Science (11JGB057) and Beijing Information Science and Technology University Discipline and Postgraduate Education New subject Finance (5028223504), the projects that support us.

## REFERENCES

Bohui Wen. Systemic Financial Risk Measure Method Research [J]. Financial Development Research, 2010.1.

Guoli Zhang. Hui Zhang. Basic Fuzzy Math and Application. Qian Kong, Chemical Industry Press, (2011).

Juan Lai. Measure of Financial Systemic Risk Based on Financial Stress Index [J]. Jianglin Lv. Statistics and Decision, No. 19, 2010.

Xiaopu Zhang. Systemic financial risk: evolution, causes, and regulatory [J]. Studies of International Finance, 2010.7.

Xu Zhao. Zhigao Xin. Enterprise Risk Warning and Management Model [J]. Journal of Beijing University of Posts and Telecommunications (Social Sciences Edition) Vol 12, No. 1, Feb. 2010.

*Information Systems and Computing Technology – Zhang & Gu (eds)*
*© 2013 Taylor & Francis Group, London, ISBN 978-1-138-00115-2*

# Research on smart mobile phone user experience with grounded theory

J.P. Wan & Y.H. Zhu
*School of Business Administration, South China University of Technology, Guanzhou, P.R. China*

ABSTRACT: The user experience influencing factors of smart mobile phone are explored in order to assess its quality. At first, we discover user experience influencing factors of smart mobile phone and establish user experience quality assessment model with grounded theory, which including both environmental experience and user experience, then calculated the weight for each factor with analytic hierarchy process: interaction (0.115), usability (0.283), durability (0.091), innovation (0.104), screen vision (0.098), appearance design (0.071), touch experience (0.057), entertainment (0.133), emotional beggar (0.048), and first four key influencing factors are discovered as follow: usability, entertainment, interaction and innovation. Finally, the model is verified through quality assessment for five smart mobile phone, we hope that it can give some inspiration to mobile phone manufacturers and operators.

## 1 INTRODUCTION

The Harvard Business Review ran an article titled "Welcome to the Experience Economy". Pine and Gilmore, who wrote a book by the same name, argue that the entire history of economic progress could be captured as a progression from extracting commodities (agrarian economy), to making goods (industrial economy), to delivering services (service economy), and now, to one of staging experiences (experience economy). The article suggests characteristics of desirable experiences that draw heavily from entertainment and customer service, as well as five principles for designing such experiences: theme the experience, fulfill it in all the details, harmonize the impression with positive cues, eliminate negative cues, and mix in memorabilia (Pine, B.J. II and Gilmore, J.H. 1998).

This paper is organized as follow: introduction and literature review are in the first, then research design, discovery user experience influencing factors of smart mobile phone with grounded theory, and establish user experience quality assessment model for smart mobile phone are in the following, quality assessment for five smart mobile phone and conclusions are in the end.

## 2 LITERATURE REVIEW

Mohammed defined the user experience as the motivators and feedback feeling during the interaction with the product or the website (Mohammed R.A. 2002). Jesse J.Garrett thought it is the performance and operation of the product in the real world (Garrett J.J. 2002). Arhippainen said "user experience is the user's immediate emotion and expectation under interaction with the machine" (Arhippainen L. 2009). J.P. Wan and H. Zhang once studied on influencing factors of Web 3D user experience with grounded theory and Analytic Hierarchy Process (AHP), and discovered four influencing factors: website quality, external environmental factors, user internal factors and recommendation, and the influencing factor model of Web 3D user experience was tried to be established (Jiangping Wan, Hui Zhang, 2011). Ground theory to provide the basic knowledge and procedures needed by persons who are about to embark on their first qualitative research projects and who want to build theory at the substantive level (Corbin, J. & Strauss, 1990). The main steps of grounded theory are as follows: (1) Theoretical sampling. (2) Collecting information; (3) Coding information, and forming the concepts from information. (4) Continually comparing between data, and between conceptions and between data and conceptions. (5) Forming theoretical conceptions, and establishing the relationships between conceptions. (6) Building theory and judging it.

## 3 RESEARCH DESIGN

Research framework is illustrated in Figure 1. At first, we discover user experience (UX) influencing factors of smart mobile phone with grounded theory, and establish user experience quality assessment model for smart mobile phone with AHP and discover the four influencing factors. Finally, there are quality assessment for five smart mobile phone and conclusions.

## 4 DISCOVER USER EXPERIENCE INFLUENCING FACTORS OF SMART MOBILE PHONE WITH GROUNDED THEORY

The idea of the user is understood with questionnaires and interviews, and the user experience's influencing factors are identified with grounded theory.

The characteristics of the 20 interviewed undergraduates as follow: (1) higher-level education, accept and understand new things quickly; (2) using smart phone more than two years, familiar with at least five kinds of smart phones; (3) loving and enjoying life.

We discover 12 categories and 60 concepts through theoretical sampling, collecting information, coding information, continually comparing between data and conceptions (Table 1).

The key task of axial coding is to explore and establish a relationship between the categories, rather than build a comprehensive theoretical model. We obtain 2 categories and 4 subcategories with axial coding (Table 2).

The core categories "User experience" were identified through continually comparing between primary sources of information, and between the results of the open coding and between the results of the axial coding (Fig. 2).

User experience influencing factors of smart mobile phone were discovered with grounded theory, mainly included the environment experience and user experience. External environment included the signal, network, virus, Economic factors, and other external factors. The three dimensions of user experience were service quality, sensory quality and emotion quality. Service quality included interaction, usability, durability and innovation; Sensory quality included screen vision, appearance design and touch experience; emotion quality included entertainment and emotional beggar.

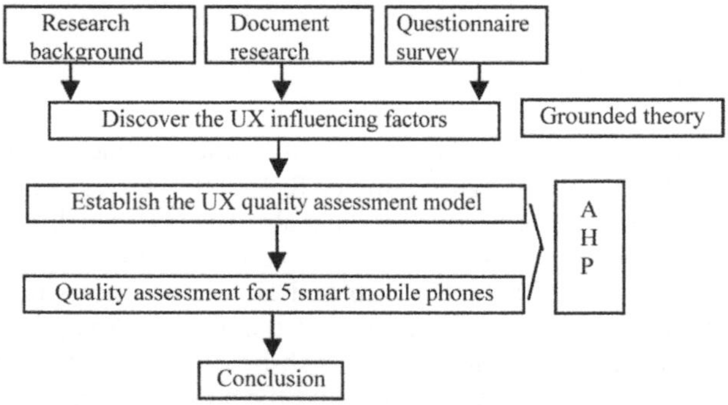


Figure 1. Research framework.

Table 1. The result of the open coding.

| Category | Conception |
|---|---|
| Appearance design ($A_1$) | Thickness; weight; screen size; appearance; color |
| Screen vision ($A_2$) | Eye-friendly; resolution ratio; blank screen; exchange theme freely; luminance accurate |
| Economic factors ($A_3$) | High performance; save network flow; save electric quantity; guarantee; maintain service |
| Innovation ($A_4$) | Television; data synchronism; shared file; memory function; location; anti-theft |
| Interaction ($A_5$) | Not smooth; get stuck easily; slow reaction; time-consuming starting; compatibility; data loss; input error; system halted |
| Usability ($A_6$) | Background operation; task switching; abundant resources; download and install conveniently; shortcut key; voluntary liquidations |
| Durability ($A_7$) | Anti-shock; stand-by time; waterproof; shockproof |
| Touch experience ($A_8$) | Finger paralysis; easy to heating; one-handed performance; click screen validly |
| Entertainment ($A_9$) | Double-side camera; high tone quality; high video quality; high network speed; games |
| Safety ($A_{10}$) | Mobile phone virus; formation disclosure; spam message; crank call; embedded virus |
| Application environment ($A_{11}$) | Automatic shift net; spam advertisement; bad signal |
| Emotional beggar ($A_{12}$) | Brand preference |

Table 2. The result of axial coding.

| Category | Subcategory | Main categories |
|---|---|---|
| Application environment ($A_{11}$) | External environment ($B_1$) | Environment experience ($C_1$) |
| Safety ($A_{10}$) | | |
| Economic ($A_3$) | | |
| Appearance design ($A_1$) | Sensory quality ($B_2$) | User experience ($C_2$) |
| Screen vision ($A_2$) | | |
| Touch experience ($A_8$) | | |
| Usability ($A_6$) | Service quality ($B_3$) | |
| Durability ($A_7$) | | |
| Interaction ($A_5$) | | |
| Innovation ($A_4$) | | |
| Entertainment ($A_9$) | Emotion quality ($B_4$) | |

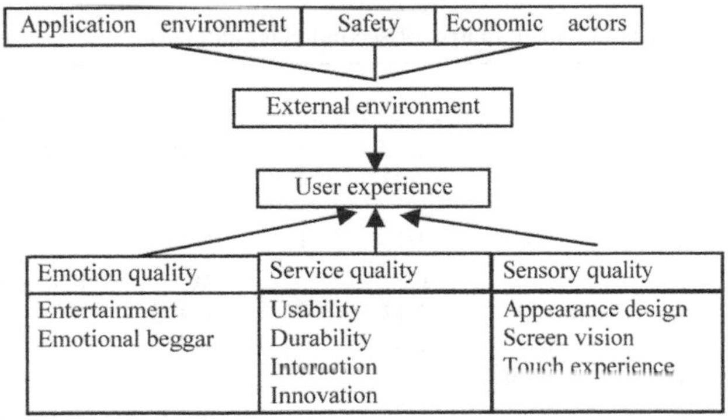


Figure 2. User experience influencing factors model.

## 5 PREFERENCES, SYMBOLS AND UNITS

Having discovery user experience influencing factors of smart mobile phone, we obtained the factor model which consists of three levels, and transformed the model to be the user experience quality assessment model for smart phone (Table 3).

We can see the ranking of the first-class indicators in Table 6 (Tables 4 and 5): $B_1$ Service quality, $B_2$ Sensory quality, $B_3$ Emotion quality; the ranking of the second-class indicators is: usability, entertainment, interaction, innovation, screen vision, durability, appearance design, touch experience, emotional beggar.

According to the ranking, we obtained the first four key influencing factors as follows: usability, entertainment, interaction and innovation.

## 6 QUALITY ASSESSMENT FOR FIVE SMART MOBILE PHONES

Basing on the IDC ranking data in January 2013 (Table 7), the Smartphone brands we selected were scale meaning table, we use the software Expert Choice to evaluate and comprehensive the result, the final matrix was illustrated in Table 8.

Table 3. User experience quality assessment hierarchy.

| Destination layer | First-class indicator | Second-class indicator |
|---|---|---|
| A User experience | $B_1$ Service quality | $C_1$ Interaction |
| | | $C_2$ Usability |
| | | $C_3$ Durability |
| | | $C_4$ Innovation |
| | $B_2$ Sensory quality | $C_5$ Screen vision |
| | | $C_6$ Appearance design |
| | | $C_7$ Touch experience |
| | $B_3$ Emotion quality | $C_8$ Entertainment |
| | | $C_9$ Emotional beggar |

Table 4. The judgment matrix of first-class indicator.

| A | $B_1$ | $B_2$ | $B_3$ | $W_i$ | |
|---|---|---|---|---|---|
| $B_1$ | 1 | 1/3.2 | 3.89 | 0.593 | C.R. = 0.01 |
| $B_2$ | 3.2 | 1 | 1.34 | 0.226 | |
| $B_3$ | 1/3.89 | 1/1.34 | 1 | 0.181 | |

Table 5. The judgment matrix of second-class indicator.

| $B_1$ | $C_1$ | $C_2$ | $C_3$ | $C_4$ | $W_i$ | |
|---|---|---|---|---|---|---|
| $C_1$ | 1 | 1/4.8 | 1/1.8 | 1 | 0.194 | C.R. = 0.02 |
| $C_2$ | 4.8 | 1 | 2.25 | 4.2 | 0.478 | |
| $C_3$ | 1.8 | 1/2.25 | 1 | 1.4 | 0.153 | |
| $C_4$ | 1 | 1/4.2 | 1/1.4 | 1 | 0.175 | |
| $B_2$ | $C_5$ | $C_6$ | $C_7$ | – | $W_i$ | C.R. = 0.03 |
| $C_5$ | 1 | 1.8 | 2.13 | – | 0.435 | |
| $C_6$ | 1/1.8 | 1 | 1.65 | – | 0.313 | |
| $C_7$ | 1/2.13 | 1/1.65 | 1 | – | 0.252 | |
| $B_3$ | $C_8$ | $C_9$ | – | – | $W_i$ | C.R. = 0.04 |
| $C_8$ | 1 | 2.77 | – | – | 0.735 | |
| $C_9$ | 1/2.77 | 1 | – | – | 0.265 | |

Table 6. The weight of user experience quality assessment dictator.

| User experience assessment indicators | $B_1$ Service quality 0.593 | $B_2$ Sensory quality 0.226 | $B_3$ Emotion quality 0.181 | $W_i$ |
|---|---|---|---|---|
| $C_1$ Interaction | 0.194 | – | – | 0.115 |
| $C_2$ Usability | 0.478 | – | – | 0.283 |
| $C_3$ Durability | 0.153 | – | – | 0.091 |
| $C_4$ Innovation | 0.175 | – | – | 0.104 |
| $C_5$ Screen vision | – | 0.435 | – | 0.098 |
| $C_6$ Appearance design | – | 0.313 | – | 0.071 |
| $C_7$ Touch experience | – | 0.252 | – | 0.057 |
| $C_8$ Entertainment | – | – | 0.735 | 0.133 |
| $C_9$ Emotional beggar | – | – | 0.265 | 0.048 |

Table 7. Top five smartphone shipments, and market share.

| Vendor | Unit shipments | Market share | Year over year change |
|---|---|---|---|
| 1. Samsung | 215.8 | 30.3% | 129.1% |
| 2. Apple | 135.9 | 19.1% | 46.9% |
| 3. Nokia | 35.1 | 4.9% | –54.6% |
| 4. HTC | 32.6 | 4.6% | –25.2% |
| 5. Blackberry | 32.5 | 4.6% | –36.4% |
| Others | 260.7 | 36.5% | 92.7% |
| Total | 712.6 | 100.0% | 44.1% |

Source: IDC Worldwide Mobile Phone Tracker, January 24, 2013 (IDC, 2013).

The characteristics of the eight experts were as follows: (1) higher-level education, very familiar with the smart phones and Internet business; (2) using smart phone more than three years, familiar with the five smart phones we selected; (3) loving and enjoying life.

According to the above calculation results, the C.R. values of all the judgment matrices are less than 0.1. Therefore, all judgment matrices pass the consistency check and the results are proved reliable.

First, we calculated the weight for each factor with AHP, the results: interaction (0.115), usability (0.283), durability (0.091), innovation (0.104), screen vision (0.098), appearance design (0.071), touch experience (0.057), entertainment (0.133), emotional beggar (0.048). Second, we calculated the weight of each second-class indicator for each phone brand (Table 9), and then added them to obtain the weight of the user experience of five different phones (Table 10).

Therefore, we use the AHP to get the descending ranking of the five mobile phones: Apple, Samsung, Nokia, HTC and Blackberry. Choosing the values of nine second-class indicators of Apple: interaction (0.402), usability (0.413), durability (0.151), innovation (0.413), screen vision (0.329), appearance design (0.354), touch experience (0.288), entertainment (0.425), emotional beggar (0.452).

In the research of the user experience quality assessment model, we obtained four key influencing indicators: usability, entertainment, interaction and innovation. The values of Apple are as follow: usability (0.413), entertainment (0.425), interaction (0.402) and innovation (0.413). We can see its values are far more than the others. Apple iPhone has a obviously competitive advantage in the user experience.

What makes an iPhone unlike anything else? Maybe it's that it lets you do so many things. Or that it lets you do so many things so easily. Those are two reasons iPhone owners say they love their iPhone. But there are many others as well.

Table 8. The assessment matrix of UX for five smart mobile phone.

| | $C_1$ (0.04) | | | | | | $C_2$ (0.03) | | | | | | $C_3$ (0.05) | | | | |
|---|---|---|---|---|---|---|---|---|---|---|---|---|---|---|---|---|---|
| $C_1$ | $D_1$ | $D_2$ | $D_3$ | $D_4$ | $D_5$ | $C_2$ | $D_1$ | $D_2$ | $D_3$ | $D_4$ | $D_5$ | $C_3$ | $D_1$ | $D_2$ | $D_3$ | $D_4$ | $D_5$ |
| $D_1$ | 1 | 1/3 | 4.2 | 3.8 | 2.7 | $D_1$ | 1 | 1/2.3 | 5.0 | 3.9 | 4.9 | $D_1$ | 1 | 1/1.4 | 1/1.2 | 1/4.4 | 1/2 |
| $D_2$ | 3 | 1 | 6.1 | 4.1 | 4.5 | $D_2$ | 2.3 | 1 | 7.5 | 5.4 | 7.2 | $D_2$ | 1.4 | 1 | 1.3 | 1/2.8 | 1/1.4 |
| $D_3$ | 1/4.2 | 1/6.1 | 1 | 1/2.5 | 1/1.8 | $D_3$ | 1/5 | 1/7.5 | 1 | 1/1.8 | 1/1.1 | $D_3$ | 1.2 | 1/1.3 | 1 | 1/4.1 | 1/1.8 |
| $D_4$ | 1/3.8 | 1/4.1 | 2.5 | 1 | 1.4 | $D_4$ | 1/3.9 | 1/5.4 | 1.8 | 1 | 1.7 | $D_4$ | 4.4 | 2.8 | 4.1 | 1 | 2.3 |
| $D_5$ | 1/2.7 | 1/4.5 | 1.8 | 1/1.4 | 1 | $D_5$ | 1/4.9 | 1/7.2 | 1.1 | 1/1.7 | 1 | $D_5$ | 2 | 1.4 | 1.8 | 1/2.3 | 1 |
| | $C_4$ (0.02) | | | | | | $C_5$ (0.01) | | | | | | $C_6$ (0.03) | | | | |
| $C_4$ | $D_1$ | $D_2$ | $D_3$ | $D_4$ | $D_5$ | $C_5$ | $D_1$ | $D_2$ | $D_3$ | $D_4$ | $D_5$ | $C_6$ | $D_1$ | $D_2$ | $D_3$ | $D_4$ | $D_5$ |
| $D_1$ | 1 | 1/2.2 | 2.8 | 3.2 | 2.7 | $D_1$ | 1 | 1/1.8 | 2.4 | 2.3 | 1.9 | $D_1$ | 1 | 1/1.9 | 2.8 | 2.4 | 2.6 |
| $D_2$ | 2.2 | 1 | 1/4.2 | 1/4.9 | 1/4.0 | $D_2$ | 1.8 | 1 | 2.9 | 2.9 | 2.2 | $D_2$ | 1.9 | 1 | 3.8 | 3.2 | 3.4 |
| $D_3$ | 1/2.8 | 4.2 | 1 | 1.1 | 1 | $D_3$ | 1/2.4 | 1/2.9 | 1 | 1.1 | 1/1.2 | $D_3$ | 1/2.8 | 1/3.8 | 1 | 1/1.5 | 1/1.2 |
| $D_4$ | 1/3.2 | 4.9 | 1/1.1 | 1 | 1/1.1 | $D_4$ | 1/2.3 | 1/2.9 | 1/1.1 | 1 | 1/1.2 | $D_4$ | 1/2.4 | 1/3.2 | 1.5 | 1 | 1.2 |
| $D_5$ | 1/2.7 | 4 | 1 | 1.1 | 1 | $D_5$ | 1/1.9 | 1/2.2 | 1.2 | 1.2 | 1 | $D_5$ | 1/2.6 | 1/3.4 | 1.2 | 1/1.2 | 1 |
| | $C_7$ (0.06) | | | | | | $C_8$ (0.02) | | | | | | $C_9$ (0.05) | | | | |
| $C_7$ | $D_1$ | $D_2$ | $D_3$ | $D_4$ | $D_5$ | $C_8$ | $D_1$ | $D_2$ | $D_3$ | $D_4$ | $D_5$ | $C_9$ | $D_1$ | $D_2$ | $D_3$ | $D_4$ | $D_5$ |
| $D_1$ | 1 | 1/1.2 | 2.5 | 2.4 | 2.1 | $D_1$ | 1 | 1/1.8 | 3.8 | 4.2 | 3.5 | $D_1$ | 1 | 1/3.8 | 1.6 | 1.4 | 2.3 |
| $D_2$ | 1.2 | 1 | 2.8 | 2.7 | 2.2 | $D_2$ | 1.8 | 1 | 4.9 | 5.3 | 4.4 | $D_2$ | 3.8 | 1 | 4.3 | 4.1 | 5.2 |
| $D_3$ | 1/2.5 | 1/2.8 | 1 | 1/1.1 | 1/1.5 | $D_3$ | 1/3.8 | 1/4.9 | 1 | 1.1 | 1/1.2 | $D_3$ | 1/1.6 | 1/4.3 | 1 | 1/1.3 | 1.5 |
| $D_4$ | 1/2.4 | 1/2.7 | 1.1 | 1 | 1/1.5 | $D_4$ | 1/4.2 | 1/5.3 | 1/1.1 | 1 | 1/1.3 | $D_4$ | 1/1.4 | 1/4.1 | 1.3 | 1 | 1.8 |
| $D_5$ | 1/2.1 | 1/2.2 | 1.5 | 1.5 | 1 | $D_5$ | 1/3.5 | 1/4.4 | 1.2 | 1.3 | 1 | $D_5$ | 1/2.3 | 1/5.2 | 1/1.5 | 1/1.8 | 1 |

Table 9. The analysis result of the dictator of five smart mobile phone.

| Dictator analysis result | D1 Samsung | D2 Apple | D3 Blackberry | D4 Nokia | D5 HTC |
|---|---|---|---|---|---|
| C1 Interaction | 0.278 | 0.402 | 0.081 | 0.136 | 0.103 |
| C2 Usability | 0.256 | 0.413 | 0.092 | 0.145 | 0.094 |
| C3 Durability | 0.103 | 0.151 | 0.111 | 0.430 | 0.205 |
| C4 Innovation | 0.283 | 0.413 | 0.103 | 0.093 | 0.108 |
| C5 Screen vision | 0.246 | 0.329 | 0.135 | 0.131 | 0.159 |
| C6 Appearance design | 0.273 | 0.354 | 0.113 | 0.136 | 0.124 |
| C7 Touch experience | 0.261 | 0.288 | 0.132 | 0.139 | 0.180 |
| C8 Entertainment | 0.298 | 0.425 | 0.089 | 0.080 | 0.108 |
| C9 Emotional beggar | 0.187 | 0.452 | 0.123 | 0.142 | 0.096 |

Table 10. The weight of the UX of five smart mobile phones.

| Mobile brand | $W_i$ |
|---|---|
| Samsung | 0.250 |
| Apple | 0.372 |
| Blackberry | 0.103 |
| Nokia | 0.153 |
| HTC | 0.122 |

Every detail and every material has been meticulously considered and refined. As a result, iPhone feels substantial in your hand and perfect in your pocket.

Millions of ways to be entertained, from one trusted source. The more apps, music, movies, and TV shows you download, the more you realize there's almost no limit to what iPhone can do. The iTunes Store is the world's largest and most trusted entertainment store. Other mobile platforms have a myriad of fragmented store options, resulting in availability issues, developer frustration, and security risks.

iPhone is so easy to use thanks to iOS 6. Innovative features like Siri and FaceTime plus built-in apps make iPhone not just useful but fun. Siri, the intelligent assistant, lets you use your voice to send messages, schedule meetings, place calls, set reminders, and more. Of course, if other vendors wanted to give users a better experience and get more market share, they should learn how to seize and improve the key influencing indicators, and continuously improve and develop themselves.

## 7 CONCLUSIONS

This article combined and analyzed the Smartphone and user experience, it had a strong practical significance. First, we establish user experience quality assessment model with grounded theory, the first-class indicators is: service quality, sensory quality, emotion quality; the second-class indicators is: usability, entertainment, interaction, innovation, screen vision, durability, appearance design, touch experience, emotional beggar. Second, we applied AHP to research the user experience quality assessment model, and obtained four key influencing indicators: usability, entertainment, interaction and innovation. We also apply AHP to get the descending ranking of the five mobile phones: Apple, Samsung, Nokia, HTC and Blackberry. After analyzing the value of Apple, we can see its values are far more than the others, Apple iPhone has a obviously competitive advantage in the user experience, all of those demonstrated the correctness of the model (In spite of many associated reasons), our results can give inspiration to smart mobile phone manufacturers and operators.

## ACKNOWLEDGEMENTS

Thanks for helpful discussion with Mr. Xiao Yan, Mr. Yongchao Zhang, Mr. Shiqiu liu and Mr. Yun Bai etc.

## REFERENCES

Arhippainen L, 2009. *Capturing user experience for product design* [EB/OL]. http://www.google.com.hk/url?q=http://robertoigarza.files.wordpress.com/2009/10/art-capturing-user-experience-for-product-design-arhippainen-2003.pdf&sa=U&ei=n9OxUZOIM4a0kAXF1IDQBQ&ved=0CBsQFjAA&usg=AFQjCNFJgSjb8B-q1HUBKcRyxze1B7CHSQ.

Corbin, J. & Strauss, 1990. A Basics of Qualitative Research: Techniques and Procedures for Developing Grounded Theory. Sage, London, UK.

Garrett J.J. 2002. The Elements of User Experience: User-Centered Design for the Web. New York: AIGA New Riders Press.

IDC, 2013. *Worldwide Mobile Phone Tracker*, January 24, [EB/OL] http://www.idc.com/getdoc.jsp?containerId=prUS239164.

Jiangping Wan, Hui Zhang, 2011. Research on Influencing Factors of Web 3D User Experience with Grounded Theory. i*Business* 3 (3): 237–243.

Mohammed R.A. 2002. Internet Marketing: building advantage in the networked economy. USA: McGraw-Hill Press.

Pine, B.J. II and Gilmore, J.H. 1998. Welcome to the Experience Economy. *Harvard Business Review*, July–August, 97.

*Information Systems and Computing Technology – Zhang & Gu (eds)*


# The software reliability analysis based on Weibull distribution

Guozhong Zou & Shaoyun Song
*Schools of Information Technology, Yuxi Normal University, Yunnan, China*

Xiaoli Huang
*YuXi Agriculture Vocation-Technical College, Yunnan, China*

ABSTRACT: During the development of Management Information System (MIS), the MIS software is passed through a truncated life test and the data of its fault are collected. By applying the method of reliability statistics, a parameter iteration of maximum likelihood equation is set up MIS Software with an operation life submitting to Weibull distribution. Various formulae of reliability index are given and verified by computation.

## 1 INTRODUCTION

There are many different lifetime distributions that can be used to model reliability data. Leemis and others present a good overview of many of these distributions. In this reference, we will concentrate on the most commonly used and most widely applicable distributions for life data analysis, as outlined in the following sections.

The Weibull distribution is a general purpose reliability distribution used to model material strength, times-to-failure of electronic and mechanical components, equipment or systems. In its most general case, the two-parameter of Weibull pdf is defined by, shown in Figure 1.

$$f(t) = \frac{\beta}{\eta}\left(\frac{t}{\eta}\right)^{\beta-1} e^{-\left(\frac{t}{\eta}\right)^{\beta}}$$

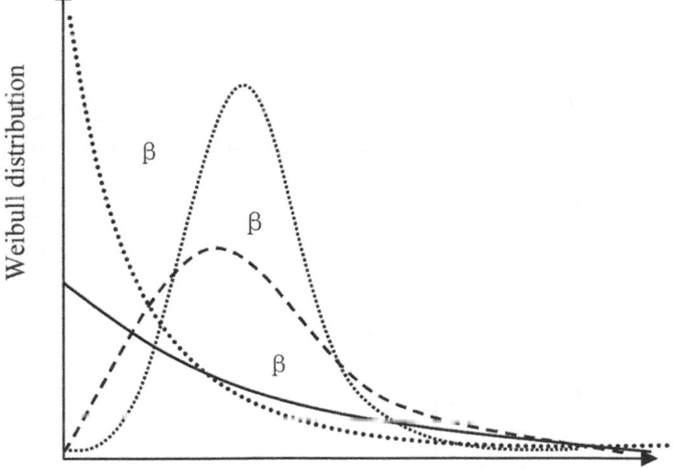


Figure 1. The two-parameter Weibull pdf.

$\beta = 1$, becomes the exponential distribution.

$\beta = 2$, for the Rayleigh distribution.

Research shows that the software project life cycle model, the project's defect removal model so well with the Rayleigh model.

With two parameters $\beta$ and $\eta$, where $\beta$ = shape parameter, $\eta$ = scale parameter.

The mean, $\overline{T}$ (called MTTF or MTBF by some authors) of the Weibull pdf is given by:

$$\overline{T} = \eta \cdot \Gamma\left(\frac{1}{\beta}+1\right)$$

where $\Gamma(1/\beta+1)$ is the gamma function evaluated at the value of $(1/\beta+1)$. The gamma function is defined as: $\Gamma(n) = \int_0^{\infty} e^{-x}x^{n-1}dx$.

## 2 PARAMETER ESTIMATION

### 2.1 *The collection process of software reliability data*

Data collection and analysis is the most important prerequisite for measuring software reliability, is directly related to the validity of any reliability measure the effectiveness of data collection, the data collection process must have an organized way. Software reliability-related data, including:

a. Defect data
b. Process data
c. Data: such as size, functionality, performance, etc.

Follow is the defect data collection.

Defect data collection often question and answer, the form of a report and payment of the problem report form requires the staff fill in the problem report form to collect and analyze the formation of statistical data. Figure 2 gives the collection process of software reliability data.

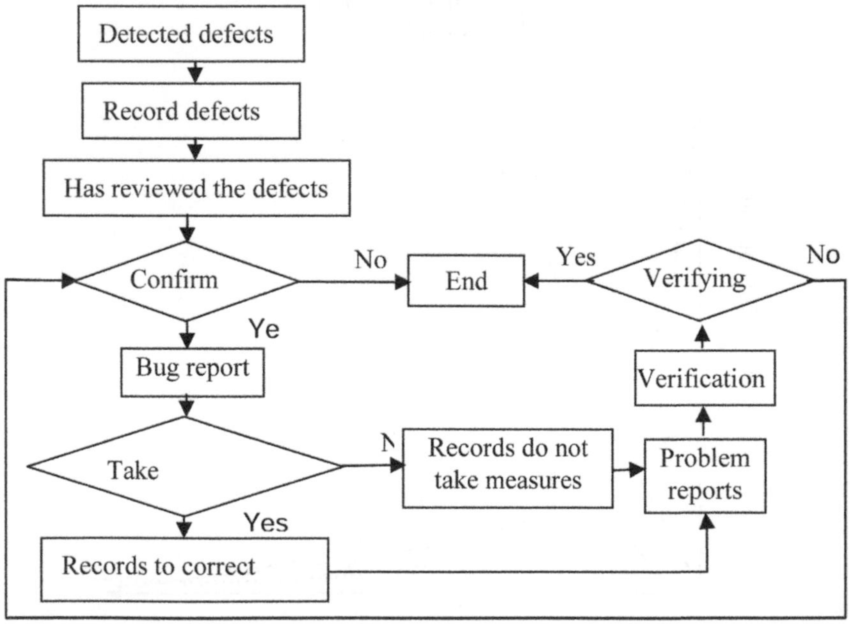


Figure 2. Collection process of software reliability data.

Follow is collection of process data.

Defect data and process integration have value, and thus the process of data collection. Usually the duration of the project as a major concern of process data, but the actual needs of more segments of data related to software reliability testing process including:

a. CPU time: nothing to do with people, but easy to ignore the work (such as assessment).
b. Calendar time: the advantage is easy to collect, but did not consider the stage characteristics, such as certain technologies are more effective at a certain stage.
c. Running time: total time to run the software for testing, did not consider the stage characteristics.
d. Other data: mean time to repair, completion of each process, the number of the software life cycle stages with the percentage of time, the amount of resources consumed by the various stages of the various stages of the closing date, the various stages of the required work to repair a defect the amount of the functional modules the number of defects, and so on.

## 2.2 *Software reliability metrics*

Software reliability $R(t)$ can be defined as: under the given conditions, at the time $[0, t]$, the probability of software failure-free operation, if $T$ is the software used to run trouble-free interval, $F(t)$ is $T$ is the cumulative distribution function, the software reliability can be expressed as:

$$R(t) = 1 - F(t)\ t \geq 0$$

Failure rate function $\lambda(t)$ as follows:

$$\lambda(t) = \lim_{\Delta t \to 0} \frac{R(t) - R(\Delta t + t)}{\Delta t R(t)} = \frac{f(t)}{R(t)}$$

Which, $f(t)$ is $F(t)$ the density function, namely:

$$f(t) = \frac{d}{dt} F(t)$$

$\lambda(t)\ \Delta t$ is the time $[0, t]$ within the software running in the $[t, t + \Delta t]$ in the conditional probability of failure can be obtained:

$$\lambda(t) = \frac{f(t)}{1 - F(t)} = \frac{d}{dt}[-\ln(1 - f(t)] = \frac{d}{dt}[-\ln R(t)]$$

Density function $f(t)$, the cumulative distribution function $F(t)$, the reliability function $R(t)$ and the failure rate function $\lambda(t)$ is closely related to the general uniquely determined by any other three, for example, if $\lambda(t)$ given, then:

$$R(t) = \exp\left\{-\int_0^t \lambda(s)ds\right\}$$

$$f(t) = \lambda(t)\exp\left\{-\int_0^t \lambda(s)ds\right\}$$

According to $f(t)$ or $R(t)$ function calculated mean time to failure MTTF, to predict the failure time

$$\text{MTTF} = \int_0^\infty tf(t)dt = \int_0^\infty R(t)dt$$

## 2.3 *Parameter estimation method*

Once a distribution has been selected, the parameters of the distribution need to be estimated. Several parameter estimation methods are available. Here introduce MLE (Maximum Likelihood) Parameter Estimation for Complete Data method.

The basic idea behind MLE is to obtain the most likely values of the parameters, for a given distribution, that will best describe the data.

As an example, consider the following data (−3, 0, 4) and assume that you are trying to estimate the mean of the data. Now, if you have to choose the most likely value for the mean from −5, 1 and 10, which one would you choose? In this case, the most likely value is 1 (given your limit on choices). Similarly, under MLE, one determines the most likely value(s) for the parameter(s) of the assumed distribution.

If $x$ is a continuous random variable with pdf: $f(x;\theta_1,\theta_2,\ldots,\theta_k)$

where $\theta_1, \theta_2, \ldots, \theta_k$ are $k$ unknown parameters which need to be estimated, with $R$ independent observations, $x_1, x_2, \ldots x_R$, which correspond in the case of life data analysis to failure times. The likelihood function is given by:

$$L(\theta_1,\theta_2,\ldots,\theta_k \mid x_1,x_2,\ldots,x_R) = L = \prod_{i=1}^{R} f(x_i;\theta_1,\theta_2,\ldots,\theta_k) \quad i=1,2,\ldots,R$$

The logarithmic likelihood function is given by: (Note: Weibull++ provides a three-dimensional plot of this log-likelihood function.

$$\ln L = \prod_{i=1}^{R} \ln f(x_i;\theta_1,\theta_2,\ldots,\theta_k)$$

The maximum likelihood estimators (or parameter values) of $\theta_1, \theta_2, \ldots, \theta_k$ are obtained by maximizing $L$.

By maximizing $\Lambda$, which is much easier to work with than $L$, the maximum likelihood estimators (*MLE*) of $\theta_1, \theta_2, \ldots, \theta_k$ are the simultaneous solutions of $k$ equations such that:

$$\frac{\partial L}{\partial \theta_i} = 0, i = 1, 2, \ldots, k$$

Even though it is common practice to plot the *MLE* solutions using median ranks (points are plotted according to median ranks and the line according to the *MLE* solutions), this is not completely representative. As can be seen from the equations above, the *MLE* method is independent of any kind of ranks. For this reason, the *MLE* solution often appears not to track the data on the probability plot. This is perfectly acceptable since the two methods are independent of each other, and in no way suggests that the solution is wrong.

# 3 SOFTWARE RELIABILITY ANALYSIS

Component refers to encapsulate data and functionality at run-time parameters can be configured by the module, usually components developed by third parties, with a clear interface description interface description in addition to other common functions, components indicates that the component should have occasion to use and a description of reliability

Table 1. The software to do the extraction.

| | | |
|---|---|---|
| Failure time $t_i$ | $t_1, t_2, \ldots, t_r$ | $t_r$ Deadline for testing |
| Number of failures | $m_1, m_2, \ldots, m_r$ | $m = m_1 + m_2 + \cdots + m_r$ |

and other performance indicators. as a basis for component-based software designer can determine the suitability of the current system components, but because the software reliability and environment-related use of the software, therefore, reliable technical indicators of expression is not just a specific target parameters, but a set of indicators parameters, can be "mapped" to represent the relationship: given a use environment, the environment returns a reliability situation. in order to use the specific environment the reliability of components under the judge, component producers need to provide any kind of data? currently no standards in this area usually consider two factors: the data must be sufficient in the specific context of software component reliability of estimates; developers have collected these data at an acceptable price range.

Component-based software design life of obedience two-parameter Weibull distribution and its density function is:

$$f(t)=\frac{\beta}{\eta}\left(\frac{t}{\eta}\right)^{\beta-1}e^{-\left(\frac{t}{\eta}\right)^{\beta}}$$

where $t>0,(\beta>0,\eta>0)$, The reliability of indicators:

The reliability function: $R_w(t)=e^{-(t/\eta)^{\beta}}$

Average life expectancy: $E_w(t)=\eta\cdot\Gamma(1/\beta+1)$

(3) failure rate: $\lambda_w(t)=\beta/\eta(t/\eta)^{\beta-1}$

The software to do the extraction at $t_i$ the truncated life test repeated the test $m_i(i = 1 \ldots r)$ (the results shown in Table 1) evaluated $\beta$, $\eta$ the maximum likelihood estimate.

In (1)(2)(3), The $\beta$ and $\eta$ using the estimated value. You can also use tools Weibull++ Version 7 to obtain these two parameters.

## 4 EXAMPLE

As the special nature of MIS software, to collect all, accurate data is very difficult life. Usually truncated life test, the failure to collect data. According D.inc incomplete data (shown in Tables I and II), for a given accuracy $\varepsilon = 0.00001$, take $\beta 0 = 1$, applying the parameter estimation method, the computer was $\beta = 1.4367$ and $\eta = 38.9396$ estimate, therefore, obey the Weibull distribution can have life the MIS software reliability indicators are:

$$R_w(t)=\exp\left(-\left(\frac{t}{38.9396}\right)^{1.4367}\right)$$

$$E_w(t)=35.3459$$

$$\lambda_w(t)=0.0369\cdot\left(\frac{t}{38.9396}\right)^{0.4367}$$

Table 2. D.inc incomplete data.

| $i$ | 0 | 1 | 2 | 3 | 4 | 5 | 6 | 7 | 8 | 9 | 10 | 11 | 12 | 13 | 14 |
|---|---|---|---|---|---|---|---|---|---|---|---|---|---|---|---|
| $t_i$ | 0 | 2 | 3 | 7 | 8 | 9 | 10 | 11 | 18 | 21 | 33 | 35 | 37 | 44 | 45 |
| $y(t)$ | 0 | 4 | 5 | 7 | 8 | 14 | 17 | 28 | 29 | 30 | 31 | 33 | 41 | 46 | 48 |
| $m_i$ | 0 | 4 | 1 | 2 | 1 | 6 | 3 | 11 | 1 | 1 | 1 | 2 | 8 | 5 | 2 |
| $i$ | 15 | 16 | 17 | 18 | 19 | 20 | 21 | 22 | 23 | 24 | 25 | 26 | 27 | 28 | 29 |
| $t_i$ | 47 | 48 | 49 | 50 | 51 | 52 | 53 | 55 | 56 | 57 | 63 | 76 | 83 | 91 | 106 |
| $y(t)$ | 50 | 53 | 56 | 59 | 64 | 67 | 68 | 69 | 71 | 74 | 76 | 78 | 79 | 80 | 81 |
| $m_i$ | 2 | 3 | 3 | 3 | 5 | 3 | 1 | 1 | 2 | 3 | 2 | 2 | 1 | 1 | 1 |

Of which: $t_i$ is the test of time (excluding holidays), $y(i)$ is the cumulative time between failures $t_i$, $m_i$ is the number $t_i$ of failures.

## 5 CONCLUSIONS

From the above derivation and application examples of the results can be seen, the use of truncated life test failure data obtained, subject to the Weibull distribution for life MIS software reliability analysis, the reliability of the software not only gives the maximum likelihood parameter index. However, equations, and iterative application of the above parameters, indicators can be calculated by the computer's maximum likelihood parameter estimate parameters of the approximate solution so as to determine the MIS software life to obey Weibull reliability index, provides an effective method.

Software quality metrics, in particular, the reliability measure, is a mature software companies, large-scale, formalized the only way, although difficult to implement, or even very difficult, but must do it. Software quality metrics is to improve the development process, and improving the CMM level, the key to improving product quality.

## REFERENCES


IEEE Std 6 10.12-1990, Glossary of Software Engineering Terminology.

Jelinski, Z., and P.B. Moranda, W. Freiberger ed., Statistical Computer Performance Evaluation, Academic Press, New York, 1972.

Leemis, Lawrence M., Reliability—Probabilistic Models and Statistical Methods, Prentice Hall, Inc., Englewood Cliffs, New Jersey, 1995.

Littlewood, B., "Software Reliability Model for Modular Program Structure", IEEE Trsans. Reliability, Vol. R-28, No. 3, Aug. 1979.

Michael, R. Lyu, Handbook of Software Reliability Engineering, Computing Mc GrawHill, New York, 1996.

Musa, J.D. A Logarithmic Poisson Execution Time Model For Software Reliability Measurement, Proc. 7th International Conf. Software Ing., 1984.

Schick, G.J., and R.W. Wolerton, "An Analysis of Competing Software Reliability Models", IEEE Trans. Software Engineering, Vol. SE-4, No. 2, Mar., 1978.

Shooman, M.L, "Structure Models for Software Reliability Prediction" Proc, International Conference of Software Engineering., IEEE Cs Press, Los Alamitos, Califolia., 1984.

Yamada, S, "S-shaped Software Reliability Growth Models and Their Applications", IEEE Trans. Reliability, Vol. R-33, No. 4, Oct. 1984.

*Information Systems and Computing Technology – Zhang & Gu (eds)*

 *ISBN 978-1-138-00115-2*

# Research on change of farmland area and its driving factors in Shaanxi, China

J.C. Han & X.M. Li

*Key Laboratory of Degraded and Unused Land Consolidation Engineering, Ministry of Land and Resources of China, Xi'an, P.R. China*
*Engineering Research Center for Land Consolidation, Shaanxi Province, Xi'an, P.R. China*

ABSTRACT: Farmland is the foundation of agriculture and food safeguard. This paper was to research the change of farmland area and its driving factors with a case study of Shaanxi province in China. By analyzing the farmland area data from 1991 to 2010, the change was researched. Driving factors including natural factors, economic factors and social factors were collected, grey incidence analysis, correlation analysis and principal component analysis were used to filter and predigest driving factors. Research results indicated total population was the most important factor which influenced the farmland area change, and farmaland area could be estimated with annual average wind speed, frostless period and total population. Results also showed driving factors were diverse in different periods, economy development and land policy were also important, so planned parenthood should be insisted, and farmland should be protected while developing economy by some measures such as promulgating land policy or laws.

## 1 INTRODUCTION

The farmaland resources is the foundation of human survival and development. Under the situation of rapid rural and urban transformation, the contradictions within farmland, economic development and environment become more and more obvious. So the farmland change and its driving factors are urgently needed to be studied (Liu et al., 2005b). The research on Land-Use/Land-Cover Change is becoming the key problem in global environment research, there is a need to analyze the driving factors for the farmland changes. Many forces have been proposed as significant (Turner II et al., 1994). Farmland area change and its driving factors have been studied with many method. F. Mialhe used an agent-based model by analyzing land use dynamics (Mialhe et al., 2012). Liu and Zheng collected the data from land survey to study the farmland change, and the driving factors were analyzed with principal component analysis and stepwise regression analysis (Liu et al., 2011; Zheng et al., 2007). Pu researched the driving forces in Xishan county with correlaiton anaysis, principal component analysis and regression analysis (Pu et al., 2002). Zhang studied the driving forces in Ansai county with typical correlaiton anaysis quantificationally, and redundancy analysis were made to test the factors (Zhang et al., 2003). Cao studied the correlation between the farmland and many factors to get the main influencing factor (Cao et al., 2008). Bai researched the dynamic of land use with system theory (Bai and Zhao, 2001). And the research by Zhang Qiu-Ju indicated the assessment of socio-economic and policy forces were the driving factors of agricultural landscape in the semiarid hilly area of the Loess Plateau, China (Zhang et al., 2004). Petr Sklenicka used general linear modeling to research Factors affecting farmland prices in the Czech Republic (Sklenick et al., 2013). Shuhao Tan analysed the factors contributing to land fragmentation (Tan et al., 2006), and Jing Wang used the data obtained from the national land surveys to analyze changes in land use and the policy dimension driving forces (Wang et al., 2012). And farmland change has also been sudied by remote sensing interpretation and GIS techinique (Xie et al., 2005; Liu et al., 2005a; Song et al., 2008; Wen et al., 2011).

Many factors have been proved to be significantly correlative with farmland area, and the factors might be correaltive inside themselves. While decision making, the factors should not be redundant, they must be reduced, and the factors with good correlations couldn't be all reserved at the same time. In this paper, grey incidence analysis was introduced to study the incidence grade between the driving factors and the farmland area, correaltion analysis and principle component analysis were used for factor filtration. And for research the driving effect of the factors quantificationally, regression analysis with stepwise method was used to establish a linear model. The contributions of the factors were revealed by path analysis.

## 2 METHODS

### 2.1 *Study area*

Shaanxi province is located in the middle drainage area of the yellow river with a semi-arid climate, the longitude is between 105°29′ E and 111°15′ E, and the latitude is between 31°42′ N and 39°35′ N. The terrain is narrow and long, the length is 870 km from south to north and 200~500 km from east to west, the area is about $20.58 \times 10^4$ $km^2$. Altiplano, plain, highland and basin are the main terrains (Su and Zhang, 2011). Shaanxi province is an agriculture province in the west of China (Zhao et al., 2011), it is also the important economy region in the west region (Zhou et al., 2011). Farmland is the important resources for local development.

### 2.2 *Data collection*

In this study, all the data were cited from the *Shaanxi statistical yearbook* from the year of 1991 to 2010. And the term "farmland" means agricultural land, or cultivated land.

The farmland area is influenced by many factors including natural factors, economic factors, and social factors. In this study, utilized land area, annual average precipitation, annual average temperature, annual average sunshine time, annual average wind speed, and the frostless period were introduced as the natural factors influencing the farmland area change. Gross National Product (GNP), GNP of the first industry, GNP of the second industry, GNP of the third industry, gross agricultural product, gross industrial product, gross wages, average salary, households consumption level, the peasants' consumption level, non-peasants' consumption level, financial revenue, financial expenditure, gross export value, gross import value, and utilized foreign captial value were introduced as the economic factors. Total population, agricultural population, the length of railway, and the length of highway were introduced as the social factors.

### 2.3 *Data analysis*

All the introduced factors were correlated with the farmland area more or less, while it's redundant of them as the driving factors for supporting the expert decision. So the grey incidence analysis was used for studying the correlation between the farmland area and the influencing factors.

Grey incidence analysis is a new mehod for factor analysis, it is not restricted by the dimension or amount of variabes, the incidence grades were used to weigh the contribution of the factors. The incidence grades include Absolute Incidence Grade (AIG), Relative Incidence Grade (RIG) and Comprehensive Incidence Grade (CIG). They indicate the simlilary of sequence line of variables, if there is more similary, the incidence grade is higer (Cao, 2007; Sun, 2007). Then the factors with better incidence grade by grey incidence analysis were selected. In this study, the factors with the first 5 incidence grade were selected by no matter AIG, RIG or CIG.

The selected factors included natural factors, economic factors and social factors. And there might be good correlations among the selected factors. So the correlation analysis and principal component analysis were used to study the correlation among the factors for factors filtration.

And to reveal the driving effect of the factors quantificationally, regression analysis with stepwise method was used, at the same time, the regression model of farmaland area could be obtained. And by path analysis, the contribution of the factors to the farmland area could be obtained.

## 3 RESULTS

### 3.1 *Farmland area change in recent 20 years*

By analyzing the collected satatistical data, the farmland area change was shown in Figure 1. It indicated the farmland area decreased from the year of 1991 to 1998 slowly, while between the year of 1998 and 2003, the farmland area decreased rapidly. And from the year of 2003 to 2006, the farmland area trended to be stable, from the year of 2006 to 2010, the farmland area trended to increase slowly.

### 3.2 *Grey incidence analysis*

Grey incidence analysis was introduced to study the incidence grade of influcencing factors (Li et al., 2010; Wang et al., 2011), the grey incidence grade indicates the similarity of the curves, and it could be calculated as the following.

$$\gamma_i = \frac{1}{N}\sum_{k=1}^{N}\left(\frac{\Delta_{(min)} + \rho\cdot\Delta_{(max)}}{\Delta_{i(k)} + \rho\cdot\Delta_{(max)}}\right) \tag{1}$$

In the equation, $\Delta_{(min)}$ means the minium difference, and $\Delta_{(max)}$ means the maximum difference, $\Delta_{i(k)}$ means the absolute difference of $x_i$ and the referencing curve, $\rho$ means the differentiation coefficient, $0 < \rho < 1$, and 0.5 ususlly.

Result of the grey incidence analysis was shown in Table 1. Absolute Incidence Grade (AIG), Relative Incidence Grade (RIG), Comprehensive Incidence Grade (CIG), were all obtained.

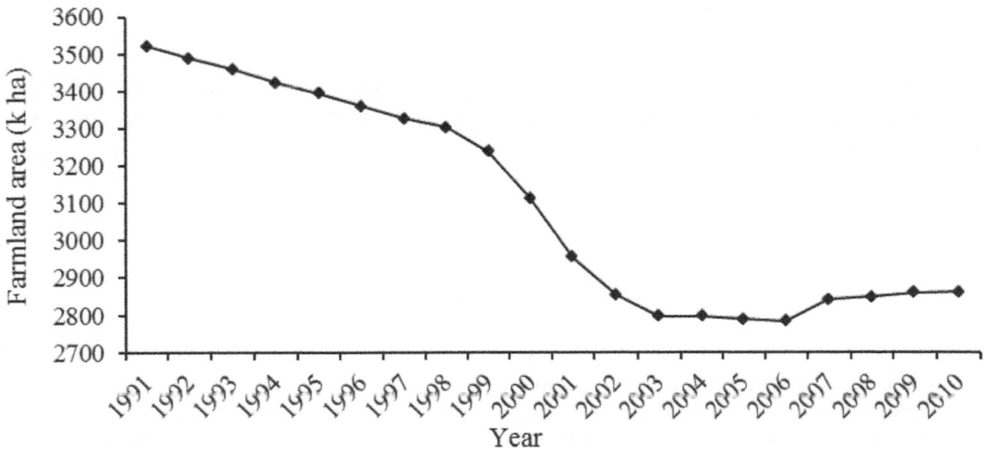


Figure 1. Change on the farmland area from 1991 to 2010.

Table 1. Result of grey incidence anlaysis.

| | $X_1$ | $X_2$ | $X_3$ | $X_4$ | $X_5$ | $X_6$ | $X_7$ | $X_8$ | $X_9$ | $X_{10}$ | $X_{11}$ | $X_{12}$ | $X_{13}$ |
|---|---|---|---|---|---|---|---|---|---|---|---|---|---|
| AIG | 0.55 | 0.67 | 0.50 | 0.51 | 0.50 | 0.54 | 0.59 | 0.79 | 0.67 | 0.73 | 0.99 | 0.58 | 0.85 |
| RIG | 0.80 | 0.72 | 0.73 | 0.60 | 0.74 | 0.89 | 0.51 | 0.53 | 0.51 | 0.51 | 0.53 | 0.51 | 0.52 |
| CIG | 0.67 | 0.70 | 0.61 | 0.55 | 0.62 | 0.72 | 0.55 | 0.66 | 0.59 | 0.62 | 0.76 | 0.55 | 0.69 |
| | $X_{14}$ | $X_{15}$ | $X_{16}$ | $X_{17}$ | $X_{18}$ | $X_{19}$ | $X_{20}$ | $X_{21}$ | $X_{22}$ | $X_{23}$ | $X_{24}$ | $X_{25}$ | $X_{26}$ |
| AIG | 0.52 | 0.64 | 0.76 | 0.56 | 0.93 | 0.96 | 0.50 | 0.50 | 0.50 | 0.79 | 0.53 | 0.66 | 0.51 |
| RIG | 0.52 | 0.53 | 0.54 | 0.53 | 0.51 | 0.51 | 0.53 | 0.51 | 0.50 | 0.84 | 0.62 | 0.63 | 0.61 |
| CIG | 0.52 | 0.58 | 0.65 | 0.55 | 0.72 | 0.74 | 0.51 | 0.51 | 0.50 | 0.81 | 0.57 | 0.64 | 0.56 |

*In the table, $X_1$ means the utilized land area, $X_2$ means the annual average precipitation, $X_3$ means the annual average temperature, $X_4$ means the annual average sunshine time, $X_5$ means the annual average wind speed, $X_6$ means the frostless period, $X_7$ means Gross National Product (GNP), $X_8$ means GNP of the first industry, $X_9$ means GNP of the second industry, $X_{10}$ means GNP of the third industry, $X_{11}$ means gross agricultural product, $X_{12}$ means gross industrial product, $X_{13}$ means gross wages, $X_{14}$ means average salary, $X_{15}$ means the households consumption level, $X_{16}$ means the peasants' consumption level, $X_{17}$ means the non-peasants' consumption level, $X_{18}$ means the financial revenue, $X_{19}$ means the financial expenditure, $X_{20}$ means the gross export value, $X_{21}$ means the gross import value, $X_{22}$ means the utilized foreign captial value, $X_{23}$ means the total population, $X_{24}$ means the agricultural population, $X_{25}$ means the length of railway, and $X_{26}$ means the length of highway, AIG means absolute incidence grade, RIG means relative incidence grade, CIG means comprehensive incidence grade, the same as below.

The results of grey incidence analysis indicated the ralation sequence by AIG as the following.

$$X_{11} > X_{19} > X_{18} > X_{13} > X_8 > X_{23} > X_{16} > X_{10} > X_9 > X_2 > X_{25} > X_{15} > X_7 > X_{12} > X_{17} > X_1 > X_6 > X_{24} > X_{14} > X_{26} > X_4 > X_{22} > X_{21} > X_{20} > X_3 > X_5.$$

The results of grey incidence analysis indicated the ralation sequence by RIG as the following.

$$X_6 > X_{23} > X_1 > X_5 > X_3 > X_2 > X_{25} > X_{24} > X_{26} > X_4 > X_{16} > X_8 > X_{11} > X_{17} > X_{15} > X_{20} > X_{13} > X_{14} > X_{12} > X_{21} > X_7 > X_9 > X_{10} > X_{19} > X_{18} > X_{22}.$$

The results of grey incidence analysis indicated the ralation sequence by CIG as the following.

$$X_{23} > X_{11} > X_{19} > X_{18} > X_6 > X_2 > X_{13} > X_1 > X_8 > X_{16} > X_{25} > X_{10} > X_5 > X_3 > X_9 > X_{15} > X_{24} > X_{26} > X_7 > X_4 > X_{17} > X_{12} > X_{14} > X_{20} > X_{21} > X_{22}.$$

### 3.3 *Factors filtration*

Denpending on the results of grey incidence analysis, the factors with the first 5 of the sequence by AIG, RIG and CIG were selected. They were $X_1$, $X_3$, $X_5$, $X_6$, $X_8$, $X_{11}$, $X_{13}$, $X_{18}$, $X_{19}$, $X_{23}$, and they are considered as more contributive to the farmalnd area. The correlation among all the factors were obtained calculated by Eq. (2), shown in Table 2.

$$r_{ij} = \frac{\sum (x_i - \bar{x}_i)(x_j - \bar{x}_j)}{\sqrt{\sum (x_i - \bar{x}_i)^2 \cdot \sum (x_j - \bar{x}_j)^2}} \tag{2}$$

Among all the left factors, $X_1$, $X_3$, $X_5$, $X_6$ were natural factors, $X_8$, $X_{11}$, $X_{13}$, $X_{18}$, $X_{19}$ were economic factors, and $X_{23}$ was social factor.

The correlations among all the natural factors were not good, all the correlation coefficients were below 0.6, so none of the natural factors was redundant, all of them could be left.

The correlation matrix indicated all the economic factors had good correlations, all the correlation coefficients were above 0.98, so if all the factors were reserved, they were redundant.

Table 2. Correlation matrix of the factors.

| Correlation | $X_1$ | $X_3$ | $X_5$ | $X_6$ | $X_8$ | $X_{11}$ | $X_{13}$ | $X_{18}$ | $X_{19}$ | $X_{23}$ |
|---|---|---|---|---|---|---|---|---|---|---|
| $X_1$ | 1.00 | −0.55** | 0.11 | −0.20 | −0.74** | −0.74** | −0.77** | −0.72** | −0.74** | −0.86** |
| $X_3$ | | 1.00 | 0.22 | 0.21 | 0.42 | 0.42 | 0.41 | 0.36 | 0.35 | 0.70** |
| $X_5$ | | | 1.00 | 0.21 | 0.26 | 0.27 | 0.20 | 0.20 | 0.17 | 0.12 |
| $X_6$ | | | | 1.00 | 0.15 | 0.15 | 0.15 | 0.12 | 0.11 | 0.31 |
| $X_8$ | | | | | 1.00 | 1.00** | 0.99** | 0.99** | 0.99** | 0.77** |
| $X_{11}$ | | | | | | 1.00 | 0.99** | 0.99** | 0.98** | 0.77** |
| $X_{13}$ | | | | | | | 1.00 | 0.99** | 0.99** | 0.77** |
| $X_{18}$ | | | | | | | | 1.00 | 1.00** | 0.70** |
| $X_{19}$ | | | | | | | | | 1.00 | 0.71** |
| $X_{23}$ | | | | | | | | | | 1.00 |

*In the table, **means $p < 0.01$, the same as below.

Table 3. The extraction of economic characteristics with factors.

| | Extraction |
|---|---|
| $X_8$ | 99.32% |
| $X_{11}$ | 99.21% |
| $X_{13}$ | 99.58% |
| $X_{18}$ | 99.30% |
| $X_{19}$ | 99.13% |

And factor analysis was operated with principal component analysis method for data reduction. The result showed all the economic factors could be the principal compent solely to stand for the economic condition, they all could explain more than 99% characteristics, Table 3, while $X_{13}$ has the most extraction (99.58%), then $X_{13}$ could be selected as the delegate of economic condition.

And there is only one social factial, that was $X_{23}$, it should be reserved. Then by factors filtration only the variables of $X_1$, $X_3$, $X_5$, $X_6$, $X_{13}$ and $X_{23}$ were reserved as the driving factors or indicator factors.

### 3.4 *The linear model by regression analysis*

For study the realation between the farmland area and the driving factors quantificationally, regression analysis with stepwise method (Chen et al., 2005) was introduced.

By the regression analysis, the factors of $X_1$, $X_3$ and $X_{13}$ were excluded with the method of stepwise, only the factor of $X_5$, $X_6$, and $X_{23}$ were reserved in the lineral model Eq. (3). The result showed the correlation between Y and $X_5$, $X_6$, $X_{23}$ was very good with a determination coefficient of 0.896, and it was significant at the level of 0.01. So it could be conluded that annual average wind speed, frostless period and total population could be used to estimate the farmland area quantificationally.

$$Y = 365.545 \times X_5 - 0.699 \times X_6 - 2.269 \times X_{23} + 10834.612 \ (r^2 = 0.896^{**}, n = 20) \quad (3)$$

And the contribution of the driving factors were obtained by path analysis. The contribution includes direct contribution indicated by $P_{iy}$ and indirect contribution indicated by $P_{ij}$, the direct contribution means how much the factor contributes to the independent variable which is the farmaland area in this paper, and the indirect contribution means how much the factor $X_i$ contributes to the farmaland area through the factor $X_j$, and it could be calculated

Table 4. Contributions of the driving factors.

| Path coefficient | Direct effect $P_{iy}$ | Indirect effect $P_{ij}$ | | |
|---|---|---|---|---|
| | | $\rightarrow X_5$ | $\rightarrow X_6$ | $\rightarrow X_{23}$ |
| $X_5$ | 0.1745 | | –0.0307 | –0.1091 |
| $X_6$ | –0.1427 | 0.037 | | –0.2754 |
| $X_{23}$ | –0.9027 | 0.0211 | –0.0435 | |

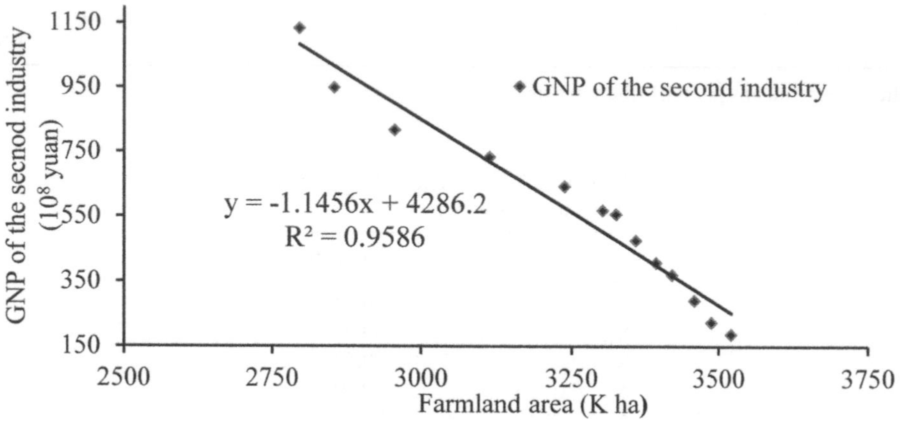


Figure 2. Correaltion between farmland area and GNP of the second industry from 1991 to 2003.

by Eq. (4). Results of path analysis were shown in Table 4, it indicated the contribution of the factors, $X_{23}$ had the most direct contribution for Y, and the natural factors of $X_5$ and $X_6$ had more indirect contribution by $X_{23}$ for Y. So it could be concluded that the population was the most important factor for farmland area change.

$$P_{ij} = r_{ij} \cdot P_{jy} \tag{4}$$

## 4 CONCLUSIONS

Research results showed the farmland area trended to decrease firstly and then increase slowly in recent 20 years. Farmland area change suffered from many factors including natural factors, economic factors and social factors. The natural factors such as temperature, precipitation and ecology protection influenced the agriculture production and the land use. The economic development was contributed by the farmland use in a certain extent, e.g., the farmland might be used to build factory, then the economic develops, while the farmalnd area decreases. And at the same time with the development of economy, the workers' wages increase, and the consumption level rised, so lots of people moved to city from village. And social factors, such as the population and the land policy could influce the farmland by farmaland assignment or land use control.

By the research result, total polulation was the first driving factor for the farmland area change, the population incresasement bring much pressure for land use. In fact, the driving factors wasn't the same in different period. From the year of 1991 to 2003, the farmland area trended to decrease, while the Gross National Product (GNP) trended to increse. With the economy development, the gross product value of the same land was different with different land use ways, so the land use transformed from low value to high value, and lots of farmland were used for non-farming land (Song et al., 2008).

## 5 DISCUSSIONS

By analyzing the correlation between the farmland area and economic factors, it could be found that the negative correlation between farmaland area and the GNP of the second was much good shown in Figure 2. Then we might conclude lots of farmlands were used to develop the second industry, the land might be used to bulid factory, marketplace and so on. And from the year of 2003 to 2010, under the situation of farmland loosing, farmland protection was paid much attention. Lots of efforts were paid for farmland increasement. And the land policy of "compensation of farmland occupancy" (Yang, 2001) was proposed and implemented. Land consolidation was carried out in a large scale, then the farmland area trended to increse stably.

In general, it could be concluded that some measures should be taken or insisted for the sustainable use of farmland. First, planned parenthood should be insisted to control the increasing spped of population since total population was the first driving factor. Second, "compensation of farmland occupancy" should be insisted, economy development should not take the cost of farmaland occupancy. The third, not the last, land consolidation of degraded and unused land should be operated to increse the amount of farmland and promote the farmland quality.

## ACKNOWLEDGEMENTS

This research was funded by the Natural Science Foundation Project of Shaanxi Province (2012JQ5015).

## REFERENCES

Bai, W.Q. & Zhao, S.D. 2001. An analysis on driving force system of land use changes. *Resources Science* 23(3): 39–41.

Cao, M.X. 2007. Research on grey incidence analysis model and its application. *Nanjing University of Aeronautics and Astronautics*, Nanjing.

Cao, Y.G., Yuan, C., Zhou, W., Tao, J. & Hua, R. 2008. Analysis on dr iving for ces and provincial differences of cultivated land change in China. *China Land Science* 22(2): 17–22.

Chen, B.F., Bai, R.P. & Liu, G.L. 2005. Stepwise regression analysis on the influencing factors of Shanxi agriculture mechanization level. *Journal of China Agricultural University* 10(4): 115–118.

Li, Q., Liang, Z.S., Dong, J.E., Fu, L.L. & Jiang, C.Z. 2010. Grey correlation formain climatic factors and quality of danshen (*Salviamiltiorrhiza* Bunge). *Acta Ecologica Sinica* 30(10): 2569–2575.

Liu, X.H., Wang, J.F., Liu, J.Y., Liu, M.L. & Meng, B. 2005b. Quantitative analysis approaches to the driving forces of cultivated land changes on a national scale. *Transactions of the CSAE* 21(4): 56–60.

Liu, X.H., Wang, J.F., Liu, M.L. & Meng, B. 2005a. Regional research on driving factors of farmland change in China. *Science in China Ser.D Earth Science* 35(11): 1087–1095.

Liu, Y.T., Zhang, Y.J. & Zhao, L. 2011. Dynamic changes of cultivated land in Changchun City and their driving forces. *Scientia Geographica Sinica* 31(7): 868–873.

Mialhe, F., Becu, N. & Gunnell, Y. 2012. An agent-based model for analyzing land use dynamics in response to farmer behaviour and environmental change in the Pampanga delta (Philippines). *Agriculture, Ecosystems and Environment*, 55–69.

Pu, L.J., Zhou, F. & Peng, B.Z. 2002, Arable land use changes at county level in Yang tze River Delta A case study of Xishan City. *Journal of Nanjing University (natural sciences)* 38(6): 779–785.

Sklenick, P., Molnarova, K., Pixova, K.C. & Salek, M.E. 2013. Factors affecting farmland prices in the Czech Republic. *Land Use Policy* 30, 130–136.

Song, K.S., Liu, D.W., Wang, Z.M., Zhang, B., Jin, C., Li, F. & Liu, H.J. 2008. Land use change in sanjiang plain and its driving forces analysis since 1954. *Acta Geographica Sinica* 63(1): 93–104.
Su, Y.L. & Zhang, Y.F. 2011. Study on effects of carbon emission by land use patterns of Shaanxi Province. *Journal of Soil and Water Conservation* 25(1): 152–156.
Sun, Y.G. 2007. Research on grey incidence analysis and its application. *Nanjing University of Aeronautics and Astronautics*, Nanjing.
Tan, S., Heerink, N. & Qu, F. 2006. Land fragmentation and its driving forces in China. *Land Use Policy* 23, 272–285.
TurnerII, B.L., Meyer, W.B. & Skole, D.L. 1994. Global land-use/land-cover change: towards an integrated study. *Ambio* 23, 91–95.
Wang, J., Chen, Y., Shao, X., Zhang, Y. & Cao, Y. 2012. Land-use changes and policy dimension driving forces in China: Present, trend and future. *Land Use Policy* 29, 737–749.
Wang, S.T., Li, X.W., Men, M.X. & Xu, H. 2011. Study on the influencing factors of grain production in Hebei Province based on gray correlation degree method. *Scientia Agricultura Sinica* 44(1): 176–184.
Wen, J.Q., Pu, L.J. & Zhang, R.S. 2011. A spatial econometric analysis on differential changes and driving forces of arable land—A case study of Jiangsu Province. *Resources and Environment in the Yangtze Basin* 20(5): 628–634.
Xie, Y.C., Mei, Y., Tian, G.J. & Xing, X.R, 2005. Socio-econornic driving forces of arable land conversion: A case study of Wuxian City, China. *Global Environ Chang* 15, 238–252.
Yang, G.S. 2001. The process and driving forces of change in arable-land area in the Yangtze River Delta during the past 50 years. *Journal of Natural Resources* 16(2): 121–127.
Zhang, Q.J., Fu, B.J., Chen, L.D. & Zhao, W.W. 2003. Arable land use changes at county level in hilly area of Loess Plateau China—A case study of Ansai County. *Journal of Soil and Water Conservation* 17(4): 146–148.
Zhang, Q.J., Fu, B.J., Chen, L.D., Zhao, W.W., Yang, Q.K., Liu, G.B. & Gulinck, H., 2004. Dynamics and driving factors of agricultural landscape in the semiarid hilly area of the Loess Plateau, China. *Agr Ecosyst Environ* 103, 535–543.
Zhao, Y.H., Zhang, L.L. & Wang, X.F., 2011. Assessment and spatiotemporal difference of ecosystem services value in Shaanxi Province. *Chinese Journal of Applied Ecology* 22(10): 2662–2672.
Zhen, H.X., Tong, J.E. & Xu, Y. 2007. Spatio-temporal changes of farmland resources and their driving forces in developed areas. *Transactions of the CSAE* 23(4): 75–78.
Zhou, D.H., Li, T.S., Hasbagen & Yang, W.L., 2011. Mechanism of county level comprehensive development spatial disparities in Shaanxi Province. *Progress in Geography* 30(2): 205–214.

*Information Systems and Computing Technology – Zhang & Gu (eds)*
*© 2013 Taylor & Francis Group, London, ISBN 978-1-138-00115-2*

# Research and application on personalized recommendation based on attribute inference technique

Zhi-qiong Bu
*Guangdong Polytechnic Normal University, Guangzhou, China*

Mei-zhen Zhu
*South-Central University for Nationalities, Wuhan, China*

ABSTRACT: To solve the existing flaws in current personalized recommendation systems such as cold start, sparse data, extensibility, real time and so on, this paper proposed the property inference method. To infer the properties, artificial neural network is used to determine the corresponding properties. This method utilizes the customer's purchase records for training the relation model between customer's attributes and the chosen products by means of ANN (Artificial Neural Network), and further infers the customer's properties, and then recommends the products with the inferred properties. The experimental result shows that the recommendation algorithm based on the property inference method can solve the ordinary flaws in current personalized recommendation systems and be able to explain the reasons of the association rules to certain extend.

## 1 INTRODUCTION

With the rapid development of the Internet technologies over the last decade, the website and users on e-commerce have been inundated by the overloaded information (Anand R.J. & David, U., 2012). Under this circumstance, personalized recommendation systems emerged (Schafer, J.B. et al., 2001). This kind of systems can analyze the users' behaviors and conclude the users' preferences, therefore, can recommendate the appropriate merchandise to the user, thus, set the users free from massive amounts of information (Schafer, J.B. et al., 2001). Of course, it can inspire the users' purchasing actions and bring great economic benefits to the enterprise, therefore, has been widely used in e-commerce sites.

Currently, there exists many personalized recommendation systems, such as collaborative filtering approach (Jun, H. & Tianming, Z., 2001), content-based recommendation (Adomavicius, G. & Tuzhilin, A., 2005), association rule based recommendation and so on. Typically, all these systems suffer from the problems such as: cold start, sparse data, extensibility, real time and so on. Because all these systems are based on the correlativity (Daling, W. et al., 2001), they are the similar problems. That is, all these systems used the similarity between the merchandise or between the users to predict the next users' actions (Chen, Y.L. et al., 2008), but seldom based on the reasons why the users buy these merchandise. For examples, the purchasers of footballs are mostly males; the purchasers of cosmetics are mostly females; the purchasers of crutches are mostly the elders. So in essence, user attributes determine his or her purchasing behavior. The recommendation based on attribute would be more accurate and effective.

This paper proposes the attribute inference algorithm based with the BP neural network, and explores the personalized recommendation based on user attributes. The BP neural network is utilized to mine every user's purchase historical records, and establish the models between user attribute and products, then recommend products based on user attributes. In this paper, to infer user attributes will be conducted through two experiments (single attribute based personalized recommendation and personalized recommendation based on attribute inference). Reliability and validity of inference technique is verified by analyzing the experimental results.

## 2 ATTRIBUTE INFERENCE MODEL BASED ON NEURAL NETWORK

### 2.1 *Definition of attribute inference*

Attribute inference, just as its name implies, is to infer customers' attributes which made the fundamental influence on customers' purchasing choices.

As there exists causality between customers' purchasing behaviors and the purchased products, the research of attribute inference is conducted. In real life, according to our conventional wisdom, the purchasers of footballs are mostly males; the purchasers of cosmetics are mostly females; the purchasers of crutches are mostly the elders; the purchasers of schoolbags are mostly students and so on. In the purchasing behaviors mentioned above gender and age of customers are involved. We can infer some attributes of customers from what they have bought. That's because there exists intrinsic causality between the characteristic possessed by commodity and some attribute of customers. In essence, what determines a user's purchasing are the attributes, such as gender, age, occupation and so on.

Therefore, we can infer a customer's attributes according to users' purchased products.

### 2.2 *Neural network*

The inference technique based on user attribute proposed in this paper needs to satisfy such features: ability to find optimal solution rapidly, high ability of self-learning and excellent ability for prediction. Among large amounts of algorithms, neural network is qualified with above requirements. In this paper, the BP neural network based model is constructed to perform the inference of customers' attributes.

Considering the advantages of the BP neural network (Rumelhart, D.E. et al., 1986), it is utilized during the process of learning and training. The BP neural network is constituted by information to dissemination and erroneous backward propagation. In the initial phase, expected outputs can be derived according to a value to which is randomly assigned. Firstly, message enter network from the 'input layer' of the network and arrive at the 'output layer' through the process of 'hidden layer'. The working process of BP neural network is shown as Figure 1. If the actual output value disagrees with desired output, errors for every output neurons would be calculated and backward propagation begins. Errors of each neuron enter the network again and then spread to every neuron of each layer in order to adjust weights of the whole network. The process will be repeated again until the termination condition is satisfied.

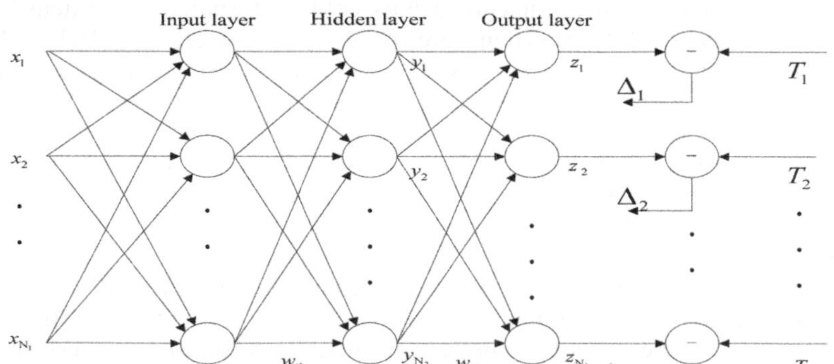


Figure 1. The working process of BP network.

## 3 EXPERIMENTS SCHEME

Data adopted in this paper come from the site of data sets (http://www.grouplens.org). In order to test the feasibility of attribute inference model based on neural network, the film genres are inferred according to customer's attribute firstly, and then recommend the type of movie according to the custom's attribute inferred by used of attribute inference. Finally, the validity and reliability of the method proposed in this paper can be verified by comparing these two recommendation results. The whole process of this experiment can be divided into four parts: 1. Attribute inference; 2. Personalized recommendation based on single attribute; 3. Personalized recommendation based on attribute inference; 4. Experimental results and analysis.

### 3.1 *Attribute inference*

The preprocessed data of MovieLens are divided into two parts: training data and predicted data: 80% and 20% of total data respectively. In this experiment, customers' attribute is inferred according customer's choice of movie genre and the corresponding rating. The movie rating and types of movie are classified into: (unknown | Action | Adventure | Animation | Children's | Comedy | Crime | Documentary | Drama | Fantasy | Film-Noir | Horror | Musical | Mystery | Romance | Sci-Fi | Thriller | War | Western), which would be taken as the input for training the neural network model and output the inferred customer's gender.

In the rest 20% predicted data, rating and genre are listed as input to predict gender according to BP neural network model proposed in this paper and then 1326 predictive values are generated as described in Table 1. In Table 1, the columns 'Genre' and 'Rating' are inputs of BP neural network; column 'Gender' is the inferred gender; Expression means the accuracy of inferred gender compared with the actual gender.

This article adopts Predict Probability to compare predicted value and actual value of gender, and then the accuracy of 68.02% is acquired.

### 3.2 *Single attribute based personalized recommendation*

According to the corresponding customer's gender stored in initial data, single attribute based personalized recommendation recommend the movie genre to customers. It takes gender and movie genre as inputs, and uses BP neural network to predict the movie rating. Value of 1 or 5 is taken to serve as the users' evaluation of each type.1 means dislike, 5 means like. Some of

Table 1. Predicted gender of customers.

| User id | Genre | Rating | Gender | Expression |
|---|---|---|---|---|
| 849 | Action | 5 | M | 0.66868430837 |
| 849 | Adventure | 5 | M | 0.67582173881 |
| 849 | Animation | 5 | M | 0.90414336183 |
| 849 | Children's | 5 | M | 0.53625865530 |
| 849 | Drama | 5 | F | 0.73444205317 |
| 849 | Horror | 5 | M | 0.67644624212 |
| 849 | Musical | 5 | M | 0.67644624212 |
| 849 | Romance | 5 | M | 0.67644624212 |
| 849 | Sci-Fi | 5 | M | 0.67644624212 |
| 849 | Thriller | 5 | M | 0.67644624212 |
| 849 | War | 5 | M | 0.67644624212 |
| 850 | Action | 5 | M | 0.66868430837 |
| 850 | Adventure | 5 | M | 0.67582173881 |
| 850 | Animation | 5 | M | 0.90414336183 |
| 850 | Children's | 5 | M | 0.53625865530 |
| 850 | Comedy | 5 | M | 0.67924078621 |
| 850 | Crime | 1 | M | 0.89537428370 |
| 850 | Drama | 1 | F | 0.69227824997 |

the results are list in Table 2. In Table 2, 'Expression' means the accuracy of predicted rating compared with the actual rating.

By the use of the neural network model proposed in this paper, the value of rating can be predicted with the input of gender and genre. In this experiment, 1020 predicted values are generated and the accuracy of 44.87% is acquired by comparing predicted values and actual values.

### 3.3 *Attribute inference based personalized recommendation*

The process of attribute inference based personalized recommendation is similar with single-attribute personalized recommendation. It also performs the personalized recommendation for the type of genre according to the corresponding customer's gender. But what make them different is that a customer's gender is inferred. The process of attribute inference is shown as Figure 2.

Table 2. Predicted value of rating with initial gender data.

| User id | Gender | Genre | Rating | Expression |
|---|---|---|---|---|
| 849 | F | Action | 5 | 0.52484003038 |
| 849 | F | Adventure | 5 | 0.65903807447 |
| 849 | F | Animation | 5 | 0.56059111056 |
| 849 | F | Children's | 1 | 0.67529153548 |
| 849 | F | Drama | 1 | 0.79750174252 |
| 849 | F | Horror | 5 | 0.53439373965 |
| 849 | F | Musical | 5 | 0.53439373965 |
| 849 | F | Romance | 5 | 0.53439373965 |
| 849 | F | Sci-Fi | 5 | 0.53439373965 |
| 849 | F | Thriller | 5 | 0.53439373965 |
| 849 | F | War | 5 | 0.53439373965 |
| 850 | M | Action | 1 | 0.54260290447 |
| 850 | M | Adventure | 5 | 0.57329498809 |
| 850 | M | Animation | 1 | 0.53916227944 |

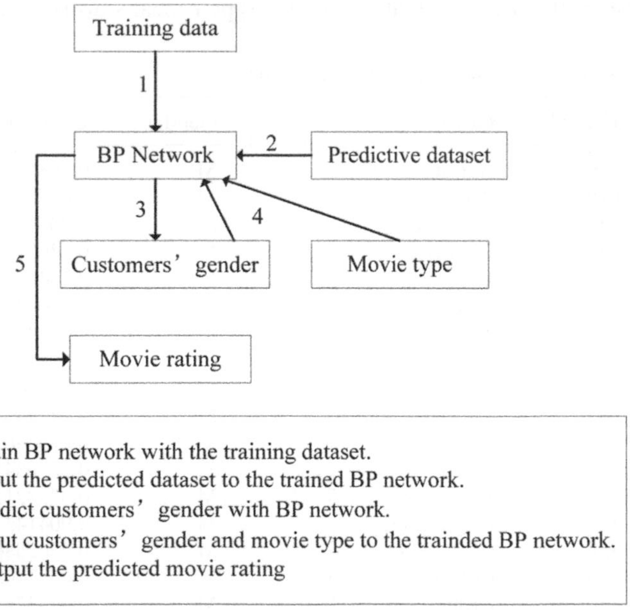
Training data
1
BP Network
2
Predictive dataset
3
4
5
Customers' gender
Movie type
Movie rating
1. Train BP network with the training dataset.
2. Input the predicted dataset to the trained BP network.
3. Predict customers' gender with BP network.
4. Input customers' gender and movie type to the trainded BP network.
5. Output the predicted movie rating

Figure 2. Working process of attribute inference.

By means of the neural network model proposed in this paper, the value of rating can be predicted with the input of gender and genre, among which the gender is predicted results from the stage of attribute inference. In this experiment, 1020 predicted values are generated and the accuracy of 19.92% is acquired by comparing predicted values and actual values. Some of the results are list in Table 3. Here, 'Expression' means the accuracy of predicted rating compared with the actual rating.

### 3.4 *Experimental results and analysis*

Single attribute based personalized recommendation utilizes customers' gender value stored in initial data as experimental data, while the customers' gender value adopt in attribute inference based personalized recommendation are derived from attribute inference. The two methods adopt the same neural network model with the input of gender and genre to predict customs' rating of movies. Table 4 gives the comparison of accuracy between two recommendation systems:

Analyses on above errors are described as following:

1. The data extracted in this experiment are just about value 1 and value 5, not the whole data. Almost 2/3 of data, in total of 138328, records are filtered, so there are errors within the measurements.

Even though diverse ratings for one genre by a customer are filtered by only extracting the ratings confined to 1 or 5, such case that one genre is given different values by a customer still exists. Table 5 gives an example of conflicting data.

Table 3. Predicted value of rating with inferred gender data.

| User id | Gender | Genre | Rating | Expression |
|---|---|---|---|---|
| 851 | M | Sci-Fi | 1 | 0.51007231784 |
| 851 | M | Thriller | 1 | 0.51007231784 |
| 851 | M | War | 1 | 0.51007231784 |
| 852 | M | Action | 1 | 0.54260290447 |
| 852 | M | Adventure | 5 | 0.57329498809 |
| 852 | M | Animation | 1 | 0.53916227944 |
| 852 | M | Children's | 1 | 0.74263840250 |
| 852 | M | Comedy | 5 | 0.62929724927 |
| 852 | M | Drama | 1 | 0.78982353305 |
| 852 | M | Romance | 1 | 0.51007231784 |
| 852 | M | Sci-Fi | 1 | 0.51007231784 |
| 852 | M | Thriller | 1 | 0.51007231784 |
| 852 | M | War | 1 | 0.51007231784 |
| 853 | M | Action | 1 | 0.54260290447 |
| 853 | M | Adventure | 5 | 0.57329498809 |

Table 4. Comparison between two recommendation results.

| Recommendation systems | Signal attribute based personalized recommendation | Attribute inference based personalized recommendation |
|---|---|---|
| Accuracy | 44.87% | 19.92% |

Table 5. Conflicting data.

| User id | Age | Gender | Occupation | Genre | Rating |
|---|---|---|---|---|---|
| 850 | 34 | M | Technician | Drama | 1 |
| 850 | 34 | M | Technician | Drama | 5 |

In the stage of attribute inference, genre and corresponding customers' rating are listed as input to predict gender and accuracy comes up to 68.02%, which means that attribute inference based on neural network shows better performance than random recommendation, and implies a research value for neural network based attribute inference. Single attribute based personalized recommendation utilizes customers' attributes stored in initial data to train the model, and further performs recommendation. In this experiment, the single attribute based personalized recommendation gets 44.87% accuracy under the circumstance that all customers' attribute are correct; attribute inference based personalized recommendation takes the use of the BP neural network trained in the first stage to perform recommendation. In this experiment, the attribute inference based personalized recommendation gets the accuracy of 19.92% under the circumstance that customers' attribute's accuracy is only 68.02%.

Through the comparison between the two experiments, attribute based personalized recommendation shows a satisfactory performance.

## 4 CONCLUSION

To solve the existing flaws in current personalized recommendation systems such as cold start, sparse data, extensibility, real time and so on, property inference method was explored from the perspective of causality in this paper. For that method, the BP neural network based property inference model is established. In order to test the validity of attribute inference model, comparison between the two results of single-attribute personalized recommendation and attribute-based inference are performed in this paper. The single attribute based personalized recommendation gets 44.87% accuracy under the circumstance that all customers' attribute are correct; attribute inference based personalized recommendation takes the use of the model trained in the first stage to perform recommendation. In this experiment, the accuracy of the attribute inference based personalized recommendation still comes up to 19.92% under the circumstance that customers' attribute's accuracy are only 68.02%. By comparing those two experimental results, we can find that attribute based personalized recommendation is able to sustain a good accuracy.

Causality has the feature of stability and universal applicability. Therefore if personalized recommendation is applied into user attribute and causality, the common problems can be better solved. In the future work, we will continuously focus on the research of attribute inference based personalized recommendation and modify the model to improve accuracy of recommendation.

## REFERENCES


Adomavicius, G. & Tuzhilin, A. 2005. Toward the next generation of recommender systems: A survey of the state-of-the-art and possible extensions. Knowledge and Data Engineering, IEEE Transactions on, 17(6): 734–749.

Anand, R.J. & David, U. 2012. Mining of Massive Datasets. Beijing.

Chen, Y.L. & Cheng, L.C. & Chuang, C.N. 2008. A group recommendation system with consideration of interactions among group members. *Expert Systems with Applications* 34(3): 2082–2090.

Daling, W. & Ge, Yu. & YuBin, B. 2001. An Approach of Association Rules Mining with Maximal Nonblank for Recommendation. *Journal of Software* 2004(8): 1182–1188.

Jun, H. Tianming, Z. 2001. Information Filtering Technology in Information Network. *Systems Engineering and Electronics* 2001(11): 76–79.

Rumelhart, D.E. & Hintont, G.E. & Williams, R.J. 1986. Learning representations by back-propagating errors. *Nature* 323(6088): 533–536.

Schafer, J.B. & Konstan, J.A. & Riedl, J. 2001. E-Commerce Recommendation Applications. *Data Mining and Knowledge Discovery* 5(1–2): 5–10.

*Information Systems and Computing Technology – Zhang & Gu (eds)*
*© 2013 Taylor & Francis Group, London, ISBN 978-1-138-00115-2*

# The electromagnetic propagation properties of metamaterials with double-inductance loading in unit cell

Yaning Liu, Suling Wang, Qingfeng Zhao & Nan Guo
*School of Electrical Engineering and Automation, Henan Polytechnic University, Jiaozuo, China*

ABSTRACT: A new structure of metamaterials is proposed. Different from the conventional structure, the units of the new structure have not one lumped inductor but dual lumped inductors, which are connected by microstrip line. Simulating experiments show that the electromagnetic properties of the new structure are similar to metamaterials with one inductance, which suggest the bandgap and left-hand passband also exist in the proposed structure. The two loaded inductances respectively have influence on the electromagnetic properties of metamaterials. It is found that the ratio of the two inductances has great relation with the frequency of the proposed metamaterials.

## 1 INTRODUCTION

In recent decades, there has been increasing interest in the research of composite artificial dielectric media that exhibit simultaneously negative electric permittivity $\varepsilon_r$ and magnetic permeability $\mu_r$ over a certain range of frequencies (D.R. Smith et al. 2000 & R.A. Shelby et al. 2001). Such metamaterias have been shown to exhibit a negative refractive index, which are also referred to as Left-handed Media (LHM) because the electric field $\vec{E}$, the magnetic field $\vec{H}$, and the propagation vector $\vec{k}$ form a left-handed triplet (V.G. Veselago 1968). These metamaterials are originally developed using thin wire strips and split-ring resonators (D.R. Smith et al. 2000 & R.A. Shelby et al. 2001). However, the size renders them impractical for the physical realization of microwave circuits. Based on the principle of metamaterials, a compact Transmission Line (TL) network exhibiting negative refraction is developed rapidly (A.K. Iyer et al. 2002 & G.V. Eleftheriades et al. 2002). The TL metamaterials are realized by periodically loading a conventional TL with lumped element series capacitors and shunt inductors, where the loading elements dominate the propagation characteristics. Usually every unit of TL metamaterials loads one lumped inductance and one capacitance.

In this letter, a kind of improved TL metamaterials loading multi-inductance is put forward. This structure is characterized by the multi-inductance that has two loaded inductors in unit cell connected by microstrip line, which has influence on the propagation properties of TL matamaterials and bandgap frequency. The simulation is implemented by Agilent ADS (Advanced Design System) software.

## 2 THEORY

Figure 1 shows the structure of conventional TL metamaterials according to the literature (G.V. Eleftheriades et al. 2002 & A. Grbic et al. 2002). The structure consists of a host TL medium periodically loaded with discrete lumped element $C_0$ and $L_0$. Every unit cell of the periodic structure consists of a capacitance and an inductance, which is connected by microstrip line. The loaded inductances are connected to the ground plane of print board through a metallized hole and the loaded capacitances are welded on the adjacent microstrip lines.

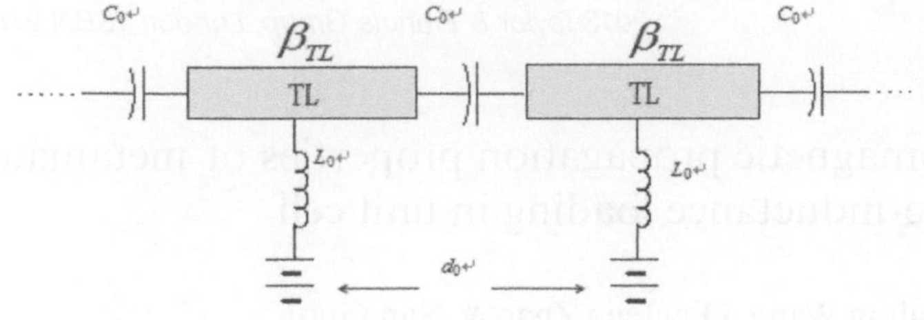


Figure 1. The structure of conventional TL metamaterials.

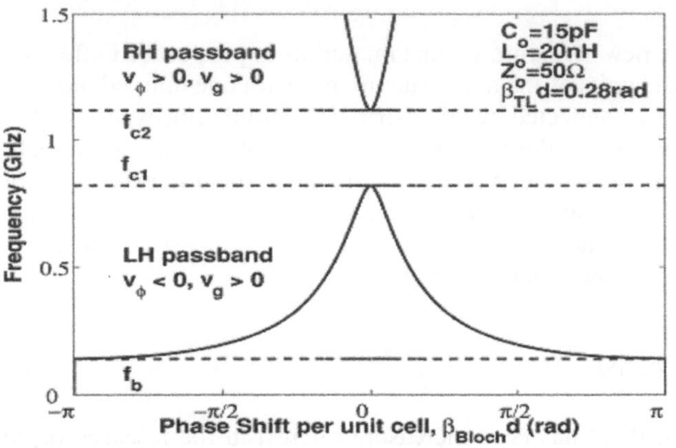


Figure 2. The dispersion relation for a unit cell with typical parameters.

The length of the unit microstrip is $d$. $L$ and $C$ are inductance and capacitance of unit microstrip. $Z_0$ means the characteristics impedance of the microstrip line. Based on the theoretical analysis of the literature (A. Grbic et al. 2002 & M.A. Antoniades et al. 2003), the dispersion equation of this structure can be written as:

$$\cos(\beta_{\text{Bloch}}d) = \cos(\beta_{\text{TL}}d)\left(1 - \frac{1}{4\omega^2 L_0 C_0}\right) + \sin(\beta_{\text{TL}}d)\left(\frac{1}{2\omega C_0 Z_0} + \frac{Z_0}{2\omega L_0}\right) - \frac{1}{4\omega^2 L_0 C_0} \tag{1}$$

where $\beta_{\text{Bloch}}$ is the Bloch transmission constant and $\beta_{\text{TL}} = \omega\sqrt{LC}$ is the propagation constant of the microstrip line.

When the wavelength of microwaves is much larger than the physical length of the unit cell ($\beta_{\text{Bloch}}d << 1$ and $\beta_{\text{TL}}d << 1$), the equivalent propagation constant can be described as (M.A. Antoniades et al. 2003)

$$\beta_{\text{eff}} \approx \pm\omega\sqrt{\left(L - \frac{1}{\omega^2 C_0 d}\right)\left(C - \frac{1}{\omega^2 L_0 d}\right)} \tag{2}$$

Figure 2 displays the dispersion relation for a unit cell with typical parameters (M.A. Antoniades et al. 2003). It can be observed that the dispersion diagram exhibits a band structure with two distinct passbands and a stopband. Expressions for the pertinent cutoff frequencies in Figure 2 are represented as follows:

$$f_b = \frac{1}{4\pi\sqrt{L_0 C_0}} \tag{3}$$

The two frequencies which correspond to the forbidden band are that

$$f_{c1} = \frac{1}{2\pi\sqrt{LC_0}}, \quad f_{c2} = \frac{1}{2\pi\sqrt{L_0 C}} \tag{4}$$

From the expressions above, the frequency of left-hand passband and bandgap frequency are closely related to the loaded capacitance and the loaded inductance. Based on the electromagnetic principle of metamaterials, a double-inductance structure metamaterial is presented in this paper.

## 3 THE PROPOSED METAMATERIALS AND SIMULATION

### 3.1 *The structure of proposed metamaterials*

Figure 3a shows the structure of double-inductor metamaterials and Figure 3b presents the unit cell of structure diagram. The unit consists of two inductances ($L_1$ and $L_2$) and one capacitance $C_0$ connected by transmission line. It is obvious that the unit in Figure 3b adds another inductance compared with Figure 1.

### 3.2 *Simulation*

Simulation is carried out by Agilent Advanced Design System (ADS) software. The characteristic impedance of the Transmission Line (TL) in Figure 3 is designed $Z_0 = 50\ \Omega$ at the frequency $f = 1.5$ GHz and the dielectric constant of the substrate in this structure is $\varepsilon_r = 4.75$. The thickness is $h = 1.5$ mm and the microstrip length is $d = 8$ mm. The lumped inductances $L_1$ and $L_2$ in Figure 3 are loaded by the holes drilled in the substrate and the lumped capacitance is set at $C_0 = 12$ pF.

Where the influence of the two inductances on the metamaterials is mainly discussed. Therefore, the simulation is carried out from two perspectives. First, the value of the capacitance is kept constant while the lumped inductance parameters $L_1$ and $L_2$ are changed. The result shows that electromagnetic bandgap of the proposed structure has changed. Then the value of the capacitance and inductance $L_1$ is kept constant, meanwhile the lumped inductance $L_2$ is changed. It is observed that there are some changes of electromagnetic bandgap of the proposed structure when the ratio of the two inductances $L_1$ and $L_2$ are accordingly varying.

Simulation 1: Two inductance values are set the same $L_1 = L_2 = L$, which the inductance value changes continuously ranging from $L = 5$ nH to 100 nH and the interval is 5 nH. Then it is observed that the forbidden bandgap has changed while the inductance value is increasing. Figure 4 shows the simulation results of typical inductances ($L$ = 5 nH, 10 nH, 15 nH). The curve with narrowest bandgap corresponds to $L = 15$ nH, the middle curve corresponds

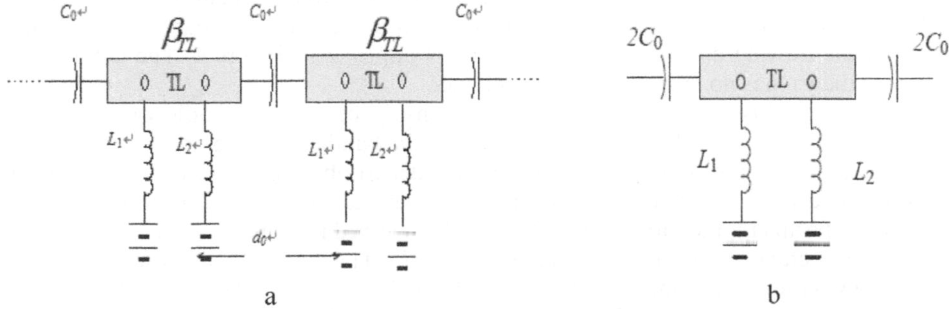


Figure 3. (a) The structure of proposed metamaterials. (b) The unit cell of structure diagram.

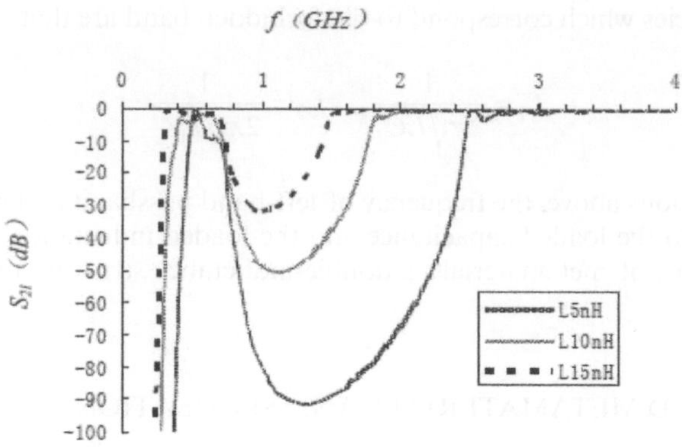


Figure 4. Forbidden band of proposed structure.

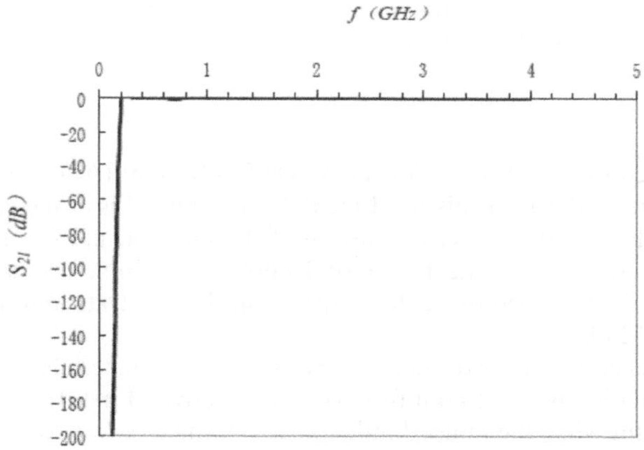


Figure 5. Bandgap disappears when inductance matching.

to $L = 10$ nH, and the widest bandgap curve corresponds to $L = 5$ nH. From Figure 4, the inductance is larger and the bandgap is smaller. It can also be found that the high frequency of bandgap moves to the left direction with the increasing value of the inductance, which is consistent with the result of expression (4).

According to Figure 4, we expect the structure could match or reach equilibrium state when the inductance is large enough. Simulation is implemented and the results have proved that the bandgap disappears when $L_1 = L_2 = 45$ nH as shown in Figure 5. Further research is also discussed: continue to increase the value of inductance from the match state, we observe that bandgap appears again and the bandgap similarly moves with the change of the inductances, but the moving direction of the bandgap is opposite to Figure 4. The detail of the simulation diagram is no longer given in this paper.

Simulation 2: This section discusses the relationship between the radio of two inductances loaded on the unit cell of metamaterials and the bandgap. There are two inductances in the unit cell. In the process of simulating, in order to observe the changes of bandgap, one inductance value is kept unchanged and the other inductance value is adjusted. In the simulation experiment, the capacitance $C_0$ is 12 pF, the fixed inductance $L_1$ is 15 nH, and the adjustable inductance $L_2$ is ranged from 5 nH to 80 nH. The typical $S_{21}$ is shown in Figure 6. The chart gives the curve of inductance $L_2 = 5$ nH, 15 nH, 30 nH, 50 nH and 80 nH. When the inductance $L_2$ is larger than 15 nH, the bandgap moves to the left direction. Meanwhile the bandgap becomes shallow, which is similar to the phenomenon that the two inductances

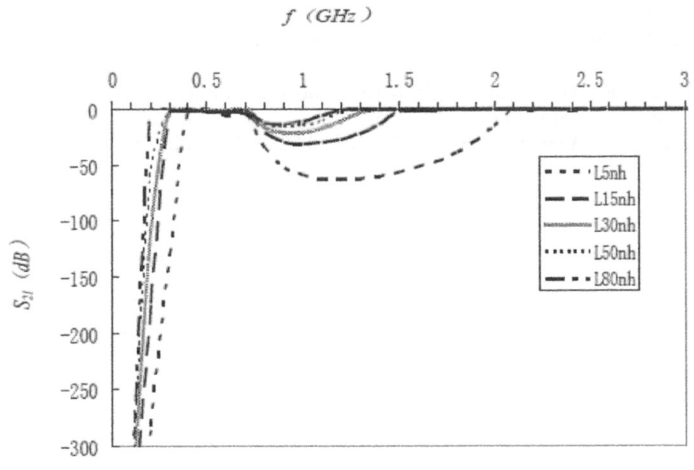


Figure 6. Bandgap changes with the ratio of the inductances.

are changing simultaneously as shown in Figure 4. However, when $L_2$ is deviated from 15 nH enough, such as the adjustable inductance reaches to 80 nH–100 nH, the bandgap no longer moves but keeps unchanged.

The phenomenon above can be explained as follows. We can consider the two inductors in the unit cell of the proposed structure are in parallel. The influence of inductances is decided by the equivalent value of the parallel inductances. When the deviation of the two inductances connected in parallel is too large, the shunt inductance value is primarily determined by the smaller one. Therefore, in the illustrated circuit, when the adjustable inductance increases to a certain value, the equivalent inductance in the unit structure will be mainly determined by the fixed inductance 15 nH and the bandgap will no longer changes.

## 4 CONCLUSION

As same as the conventional TL metamaterials, the proposed metamaterials with lumped elements loading dual-inductor structure also exist left-hand passband, bandgap and right-hand passband. When adjust the values of the two inductors are adjusted simultaneously, the forbidden band changes with the inductor. The larger value of inductors has the narrower the bandgap. When one inductance is unchanged and the other is changed, the phenomenon is slightly different: the forbidden band has close relationship to the ratio values of the two inductors. If the deviation value of the two inductors is much greater, the bandgap frequency does not change. The radio of the two inductors decides the forbidden band property of metamaterials with double-inductance loading.

## REFERENCES


Antoniades, M.A. Eleftheriades, G.V. "Compact Linear Lead/Lag Metamaterial Phase Shifters for Broadband Applications." IEEE Antennas and Propagation Letters, vol. 2, pp. 103–108, 2003.

Eleftheriades, G.V, Iyer, A.K. and Kremer, P.C. "Planar negative refractive index media using periodically L–C loaded transmission lines," IEEE Trans. Microwave Theory Tech., vol. 50, pp. 2702–2712, Dec. 2002.

Grbic, A. and Eleftheriades, G.V. "A backward-wave antenna based on negative refractive index L-C networks," in Proc. IEEE Int. Symp. Antennas and Propagation, vol. 4, San Antonio, TX, June 2002, pp. 340–343.

Iyer, A.K. and Eleftheriades, G.V. "Negative refractive index metamaterials supporting 2-D waves," in Proc. IEEE Int. Symp. Microwave Theory and Techniques, vol. 2, Seattle, WA, July 2002, pp. 1067–1070.

Shelby, R.A. Smith, D.R. and Schultz, S. "Experimental verification of a negative index of refraction," Sci., vol. 292, pp. 77–79, Apr. 2001.

Smith, D.R. Padilla, W.J. Vier, D.C. Nemat-Nasser, S.C. and Schultz, S. "Composite medium with simultaneously negative permeability and permittivity," Phys. Rev. Lett., vol. 84, pp. 4184–4187, May 2000.

Veselago, V.G. "The electrodynamics of substances with simultaneously negative values of $\varepsilon$ and $\mu$," Sov. Phys. Uspekhi, vol. 10, no. 4, pp. 509–514, Jan.–Feb. 1968.

*Information Systems and Computing Technology – Zhang & Gu (eds)*


# Online position calibration of antenna array elements for direction finding

Liang Bin Chen & Xiao Yun Zhang
*Science and Technology on Electronic Information Control Laboratory, Chengdu, China*

Qun Wan & Ji Hao Yin
*Department of Electronic Engineering, University of Electronic Science and Technology of China, Chengdu, China*

ABSTRACT: It is well known that position error of antenna array elements may deteriorate dramatically the performance of direction finding. In order to calibrate the dynamic position error, an online position calibration method is proposed. Unlike the previous offline methods dealing with only one calibrated signal located at the known direction, the proposed method can work in the presence of multiple signals with unknown directions. It is an enhanced position calibration method and ultimately takes the advantage of the subspace relationship between different signals with unknown directions. Simulation results are provided to validate the improved performance of the proposed online position calibration method.

## 1 INTRODUCTION

Sensor array and direction finding have been widely used in radar, communication, sonar, earthquake, radio astronomy, electronic surveillance and many other fields. It is an important prerequisite for finding directions of multiple signals that direction vectors of the antenna array corresponding to any possible directions in the angle range should be known in advance (Wan et al. 2013).

According to the theoretical signal model of the antenna array and given positions of antenna elements, analytical formulas can be used to determine the direction vector of the antenna array in any direction. However, it is often encountered in the practical engineering applications that the measured positions of antenna array elements are not accurate or stationary. In addition, for the different shape of the antenna array, the array position errors on the influence of the direction vector of the antenna array is not the same. In many previous calibration methods for position errors of antenna array elements (e.g., Chen et al. 2013; Backen et al. 2008; Xu et al. 2010; Jaehyun et al. 2010), the position errors of the antenna array elements were estimated by compare the difference between the analytical direction vector and measured direction vector corresponding to the same known direction. However, in the case of multiple signals, it is not easy to measure measured the direction vector corresponding a known direction.

In order to calibrate the position errors of the antenna array elements, we propose an online position calibration method. Unlike the previous offline methods dealing with only one calibrated signal located at the known direction, the proposed method can work in the presence of multiple signals with unknown directions. It is an enhanced position calibration method and ultimately takes the advantage of the subspace relationship between different signals, even their directions are unknown.

This paper is organized as follows. Section 2 briefly formulate the problem of direction finding in the presence of position error of antenna array elements. The proposed online calibration method is given in Section 3. Section 4 presents several simulation results to verify the performance of the proposed method. Section 5 provides a concluding remark to summarize the paper.

## 2 PROBLEM FORMULATION

Consider a linear array that consists of $M$ antenna. The position of the $m$-th antenna element $d_m$. Assume that the sources $s_k(t)$ come from azimuth $\theta_k$ in the far field of the array, $k = 1,2,\ldots, K$, $K$ is the number of signals. The received signal vector of the array can be expressed as:

$$\mathrm{x}(t) = \sum_{k=1}^{K} a(\theta_k)s_k(t) + v(t)$$

where $x(t)$ is the received signal vector of the array, $t$ is sampling moments, $\mathrm{v}(t)$ is the receiver noise vector, $a(\theta_k)$ is direction vector associated with azimuth $\theta_k$. The $m$-th component of $a(\theta_k)$ is

$$a_m(\theta_k) = e^{j2\pi d_m \sin(\theta_k)/\lambda}, \quad m = 1,2,\ldots, M$$

In the presence of position error of antenna array element, the actual $m$-th component of the direction vector associated with azimuth $\theta_k$ is

$$\tilde{a}_m(\theta_k) = e^{j2\pi(d_m + \varepsilon_m)\sin(\theta_k)/\lambda} \tag{1}$$

where $\varepsilon_m$ is the unknown displacement of the $m$-th antenna array element. Based on Taylor series expansion, we have the first order approximation as

$$\begin{aligned}\tilde{a}_m(\theta_k) &\approx e^{j2\pi d_m \sin(\theta_k)/\lambda} + \left(j2\pi\sin(\theta_k)/\lambda\right)e^{j2\pi d_m \sin(\theta_k)/\lambda}\varepsilon_m \\ &= a_m(\theta_k) + b_m(\theta_k)\varepsilon_m\end{aligned} \tag{2}$$

where

$$b_m(\theta_k) = \left(j2\pi\sin(\theta_k)/\lambda\right)a_m(\theta_k) \tag{3}$$

From (2), we can see that $b_m(\theta_k)$ determine the effect of position error $\varepsilon_m$. If $\theta_k$ is closer to zero, the effect of position error is smaller.

The sample autocorrelation matrix of the received vector $\mathrm{x}(t,\theta_k)$ is:

$$\mathrm{R} = \frac{1}{T}\sum_{t=1}^{T} \mathrm{x}(t)\mathrm{x}^H(t)$$

where $T$ represents the number of received signal vectors, and $[\ ]^H$ complex conjugate transposition. The singular value decomposition of the sample autocorrelation matrix is

$$R = U\Lambda U^H$$

where $\Lambda$ is a diagonal matrix whose diagonal elements correspond to the singular value of $R$, $U$ is matrix whose column vectors are singular vectors of $R$, $u_1, u_2, u_3, \ldots, u_M$, corresponding to the singular values, $\lambda_1 \geq \lambda_2 > \lambda_3 \geq \ldots \geq \lambda_M$.

According to the subspace decomposition approach, the signal subspace of the sample autocorrelation matrix is:

$$Q_n = [u_{K+1} \quad u_{L+1} \quad \dots \quad u_M] \tag{4}$$

where $K$ is the number of signals. Due to the orthogonal relationship between the noise subspace and the actual direction vector associated with azimuth $\theta_k$, we have

$$Q_n^H \tilde{a}(\theta_k) = 0 \tag{5}$$

where $\tilde{a}(\theta_k) = [\tilde{a}_1(\theta_k) \ \tilde{a}_2(\theta_k) \ \dots \ \tilde{a}_M(\theta_k)]^T$.

The problem is to estimate the azimuth $\theta_k$, $k = 1,2,\dots, K$, for all the signals in the case of unknown $\tilde{a}(\theta_k)$ due to $\varepsilon_m$, the position uncertainty of antenna array elements.

## 3 ONLINE CALIBRATION OF POSITION ERROR

### 3.1 *Analysis of identification*

Substitute (2) into (5), we have:

$$Q_n^H \left(a(\theta_k) + B(\theta_k)\varepsilon\right) = 0 \tag{6}$$

where

$$\varepsilon = [\varepsilon_1 \ \varepsilon_2 \ \dots \ \varepsilon_M]^T$$

$$a(\theta_k) = [a_1(\theta_k) \ a_2(\theta_k) \dots a_M(\theta_k)]^T$$

$$B(\theta_k) = \text{diag}\left(b(\theta_k)\right)$$

$$b(\theta_k) = [b_1(\theta_k) \ b_2(\theta_k) \ \dots \ b_M(\theta_k)]^T$$

Because the number of unknown $\varepsilon_m$ is $M-1$ (without the loss of generality, $\varepsilon_1 \equiv 0$) and the number of equations is $M-K$, even if $\theta_k$ is known, the linear equations (6) are underdetermined when $K > 1$.

However, it is noteworthy that position error $\varepsilon_m$ is a real number, we can augment (6) as

$$\begin{bmatrix} -Q_n^H B(\theta_k) \\ -Q_n^T B^*(\theta_k) \end{bmatrix} \varepsilon = \begin{bmatrix} Q_n^H a(\theta_k) \\ Q_n^T a^*(\theta_k) \end{bmatrix} \tag{7}$$

Therefore, we have $2(M-K)$ linear equations and $M-1$ unknowns, and the linear equations (7) are not underdetermined when $\theta_k$ is known and $M \geq 2K-1$.

From (7), the augmented calibration of the position error $\varepsilon$ can be estimated by

$$\varepsilon = \begin{bmatrix} -Q_n^H B(\theta_k) \\ -Q_n^T B^*(\theta_k) \end{bmatrix}^{\dagger} \begin{bmatrix} Q_n^H a(\theta_k) \\ Q_n^T a^*(\theta_k) \end{bmatrix} \tag{8}$$

where $[\ ]^{\dagger}$ represents the pseudo inverse of matrix.

As the result, the spatial spectrum for direction finding can be given by MUSIC method as

$$f(\theta) = \frac{1}{\left\| Q_n^H (a(\theta) + B(\theta)\varepsilon) \right\|_F^2} \tag{9}$$

### 3.2 *Enhanced calibration of position error*

In fact, we have $K$ sets of linear equations as (7), i.e.,

$$\begin{bmatrix} -Q_n^H \mathrm{B}(\theta_k) \\ -Q_n^T \mathrm{B}^*(\theta_k) \end{bmatrix} \varepsilon = \begin{bmatrix} Q_n^H a(\theta_k) \\ Q_n^T a^*(\theta_k) \end{bmatrix} \tag{10}$$

for $k = 1, 2, \ldots, K$. However, directions $\theta_k$ are unknown.

To take the advantage of (10), we can enhance (7) as:

$$\begin{bmatrix} -Q_n^H B(\theta_0) \\ -Q_n^T B^*(\theta_0) \\ -Q_n^H B(\theta) \\ -Q_n^T B^*(\theta) \end{bmatrix} \varepsilon = \begin{bmatrix} Q_n^H a(\theta_0) \\ Q_n^T a^*(\theta_0) \\ Q_n^H a(\theta) \\ Q_n^T a^*(\theta) \end{bmatrix} \tag{11}$$

where $\theta_0$ is a known direction of a signal, e.g., $\theta_0 = \theta_k, 1 \le k \le K$, and $\theta$ is on the grid of searching directions. Thus, the enhanced calibration of the position error $\varepsilon$ can be estimated by

$$\varepsilon = \begin{bmatrix} -Q_n^H B(\theta_0) \\ -Q_n^T B^*(\theta_0) \\ -Q_n^H B(\theta) \\ -Q_n^T B^*(\theta) \end{bmatrix}^{\dagger} \begin{bmatrix} Q_n^H a(\theta_0) \\ Q_n^T a^*(\theta_0) \\ Q_n^H a(\theta) \\ Q_n^T a^*(\theta) \end{bmatrix} \tag{12}$$

## 4 SIMULATION RESULTS

We consider a linear array of $M = 9$ antenna with half wavelength interval of the adjacent elements. Four zero-mean narrowband signals in the far-field impinge upon this array from distinct Directions Of Arrival (DOA), i.e., –34.1, –27.6, –19.3 and –10.4 degree. The number of snapshots is T = 1024, the Signal to Noise Ratio (SNR) is 10 dB. With respect to the first or reference antenna element, the unknown displacement of other antenna element is –0.1371, –0.1030, –0.1135, –0.1068, 0.0427, 0.0947, –0.1453, respectively.

In Figure 1, we compare the spatial spectrum obtained using the augmented (8) and the enhanced (12) calibration method of the position error. Direction of the first signal is

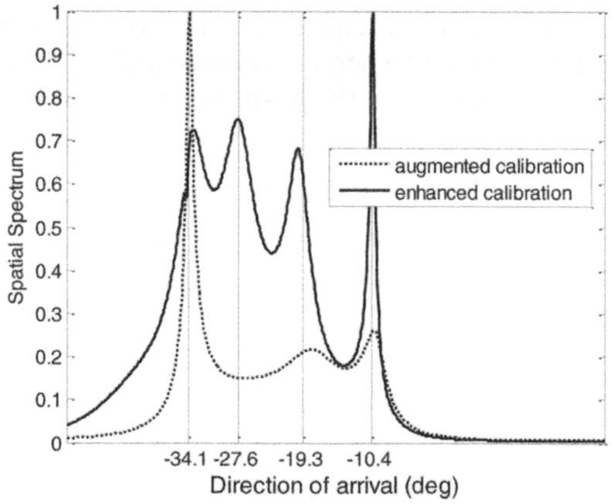


Figure 1. Spatial spectrum obtained using the augmented (8) and the enhanced (12) calibration method of the position error.

assumed to be known. It can be seen that the spatial resolution obtained by the enhanced calibration method is superior to that of the augmented method.

In Figures 2–4, we compare the spatial spectrum obtained with known position of antenna array elements, without calibration, and using the enhanced (12) calibration method of the position error. Direction of the first signal is assumed to be known. It can be seen that both the spatial resolution and direction finding accuracy obtained by the enhanced calibration method are very close to that with known position of antenna array elements. In Figure 2, we see that the effect of position error is small for the second signal because $\theta_2$ is close to zero.

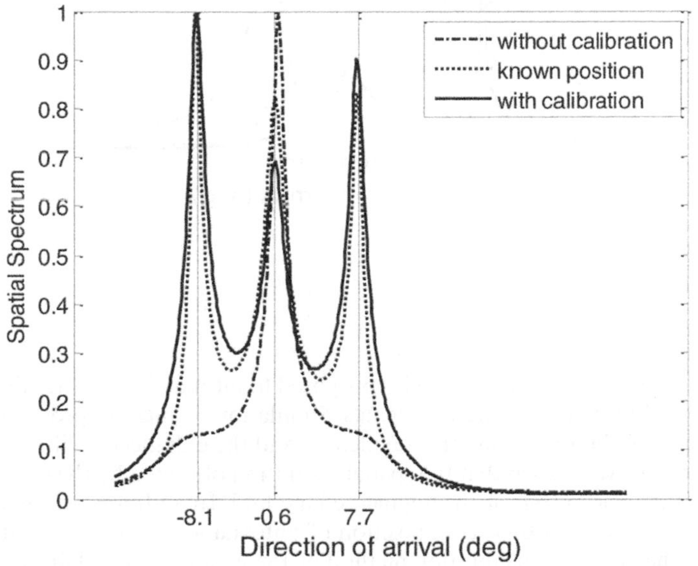


Figure 2. Spatial spectrum (three signals).

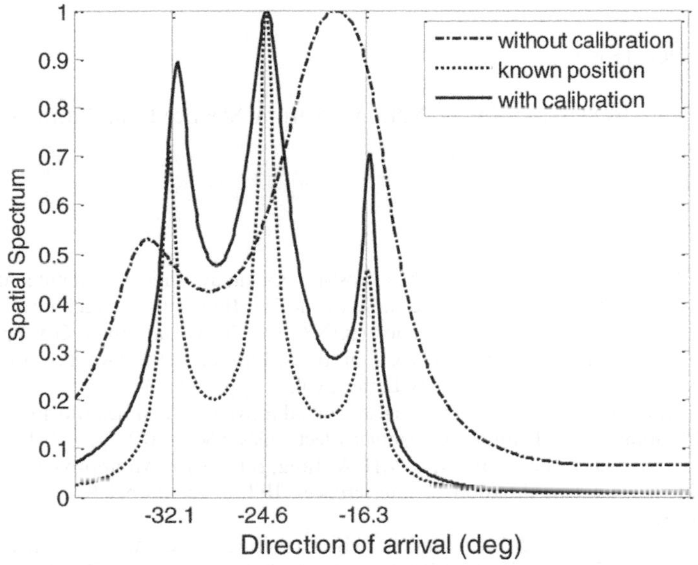


Figure 3. Spatial spectrum (three signals).

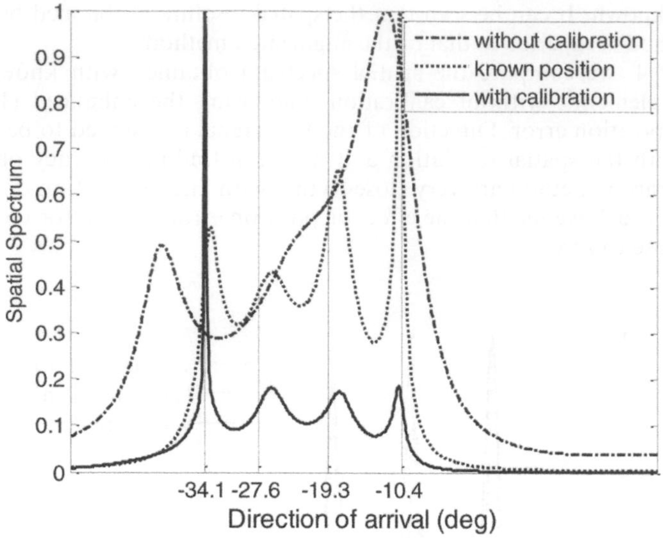


Figure 4. Spatial spectrum (four signals).

## 5 CONCLUSION

An enhanced position calibration method is proposed to ultimately take the advantage of the subspace relationship between different signals. Simulation results are provided to compare the spatial spectrum obtained using the augmented and the enhanced calibration method of the position error. It was shown that the spatial resolution obtained by the enhanced calibration method is superior to that of the augmented method. In addition, we also compare the spatial spectrum obtained with known position of antenna array elements, without calibration, and using the enhanced calibration method of the position error. Direction of the first signal is assumed to be known. It was shown that both the spatial resolution and direction finding accuracy obtained by the enhanced calibration method are very close to that with known position of antenna array elements.

## ACKNOWLEDGMENT

This work was supported in part by the National Natural Science Foundation of China under grant 61172140.

## REFERENCES

Backen, S., Akos D.M. & Nordenvaad M.L. 2008. Post-processing dynamic GNSS antenna array calibration and deterministic beamforming. Proceedings of the 21st International Technical Meeting of The Satellite Division of the Institute of Navigation (ION GNSS 2008). Savannah, GA, pp. 2806–2814.

Chen, Y., Chang, A. & Lee. H. 2000. Array Calibration Methods for Sensor Position and Pointing Errors. Microwave and Optical Technology Letters, vol. 26, pp. 132–137.

Jaehyun, P. & Joohwan, C. 2010. DTV receivers using an adaptive switched beamformer with an online-calibration algorithm. IEEE Trans. on Consumer Electronics, 56(1), 2010, pp. 34–41.

Wan, Q., Xu, B.G., Yin, J.H., Fang, F., Wan, Y.H. & Tang, S.L. 2013. An approach to manifold estimation for antenna array in situation of interferences. IET International Radar Conference 2013, Xi An, China, Apr. 14–16.

Xu, Z., Trinkle M. & Gray, D.A. 2010. A modelled eigenstructure based antenna array calibration algorithm for GPS. Proceedings of the 23rd International Technical Meeting of The Satellite Division of the Institute of Navigation (ION GNSS 2010). Portland, OR, pp. 3220–3228.

*Information Systems and Computing Technology – Zhang & Gu (eds)*
 *ISBN 978-1-138-00115-2*

# Adaptive Harris corner detection algorithm based on image edge enhancement

Yuanyuan Fang & Z. Lei
*Graduate School of Chinese Academy of Sciences, Beijing, China*
*Jiangsu R&D Center for Internet of Things, Jiangsu, China*

ABSTRACT: Harris corner detection is widely applied in the research field of image mosaic, and the calculation of the corner is based on the gradient of image pixels. In this paper, we firstly introduce the inadequacy of Harris corner detection on over-exposed images and propose an adaptive Harris corner detection method based on image edge enhancement. The new algorithm uses canny edge operator to enhance the image edge details, therefore, it solves the corners' information missing caused by over-exposure and eliminates some false corners. On the platform of the MATLAB simulation, the experimental results show that, compared with classical Harris corner detection method, the proposed algorithm avoids the effects of over-exposure and has uniform distribution of corners and less false corners.

## 1 INTRODUCTION

Edges are significant local changes of intensity in an image. Edge points are pixels where the image values undergo a sharp variation. Important features can be extracted from the edges of an image (e.g., corners, lines, curves). Edge detection is a fundamental tool in image processing and computer vision, particularly in the areas of feature detection and feature extraction, which aim at identifying points in a digital image at which the image brightness changes sharply. Edge detection significantly reduces the amount of data and filters out useless information, while preserving the important structural properties in an image. It is well known that canny edge detection algorithm is the optimal edge detector for its ability to generate single-pixel thick continuous edges.

Digital mosaic technique is useful in real life and is being used in a variety of fields. Image mosaic technique includes three important steps: feature extraction, feature matching and image fusion. The feature extraction is a key part of the whole image mosaic process, and it is directly related to stitching results. The algorithm on feature extraction can be divided into two types: one is based on contour line proposed by Kitchen L and Rosenfeld A in 1982 (Kitchen, L et al. 1982), the other is based on image gray value. The former algorithm needs to encode the image edges, and has great dependence on the edge of the image, the corner detection information will be lost if the image can not be provided completely. The latter algorithm, for example, Susan corner detection algorithm (Smith, S.M et al. 1997), Harris corner detection algorithm (Harris, C et al. 1998), Moravec corner detection algorithm (Moravec, H 1981), Sift corner detection algorithm (David G. Lowe 2004), etc., firstly compared the size of the template region's gray values with the image region's, and then matched. The accuracy of the algorithm is relatively higher, but it has some drawbacks such as complex calculations, inaccurate corner positioning and bad real-time processing of image data (Li Yi-bo et al. 2011). In order to take full advantage of the two algorithms, this paper combines the canny edge detection method with Harris corner detection algorithm, and proposes an adaptive Harris algorithm. The new algorithm enhances the image edge before extracting corners.

## 2 HARRIS CORNER DETECTION PRINCIPLE AND INADEQUACY

### 2.1 *Harris corner detection principle*

Harris corner detection algorithm is a kind of the point feature extraction algorithm based on signal point feature proposed by Chris Harris and Mike Stephens in 1988 (Stephen M. Smith et al. 1997), it is also known as the Plessey corner detection algorithm. The principle is as follows, when the image window *w* (general for rectangular area) move small displacement (*x*, *y*) in any direction, the gray variation can be defined as:

$$E_{x,y}=\sum W_{u,v}[I_{x+u,y+v}-I_{u,v}]^2=\sum_{u,v}W_{u,v}[xX+yY+o(x^2+Y^2)]^2$$
$$=Ax^2+2C(x,y)+By^2=(x,y)^{u,v}M(x,y)^T \quad (1)$$

where,

$$X=I\otimes(-1,0,1)=\frac{\partial}{\partial}$$

$$Y=I\otimes(-1,0,1)^T=\frac{\partial I}{\partial y} \quad (2)$$

$W_{u,v}$ is the coefficient of the Gaussian window at (*u*, *v*). *X* and *Y* are first order grads, which reflect the gray scale change of the elements.

### 2.2 *The inadequacy of Harris corner detection on over-exposed images*

The demands to improve the visibility quality of the captured images in extremes lighting conditions have emerged increasingly important in digital image processing. The extremes conditions are when there is lack of reasonable illumination termed as under-exposed and too much of light termed as over-exposed (Mohd Azau 2007). Over-exposure is rather annoying in photo taking. However, when irradiance across a scene varies greatly, there will always be over- or under-exposed areas in an image of the scene no matter what exposure time is used. The texture information of over-exposed image is weak. Harris corner detection algorithm is a kind of effective feature point algorithm, but it also have insufficient. Harris corner detection algorithm can extract a large number of useful corners in the area of rich texture information, but when the texture information is weak, the result is far from ideal. Questions have arisen over the detection results of over-exposed images, the number of the detected corners is too low to meet the image matching requirements.

## 3 ADAPTIVE HARRIS CORNER DETECTION ALGORITHM BASED ON CANNY EDGE DETECTION

### 3.1 *Canny edge detection principle*

The Canny edge detector is an optimal edge detection operator that uses a multi-stage algorithm to detect a wide range of edges in images. It was developed by John F. Canny in 1986 (J.F. Canny 1986). Firstly the canny edge detector smoothes the image to eliminate noise, then a simple 2-D first derivative operator is applied to the smoothed image to highlight regions of the image with high first spatial derivatives. Edges give rise to ridges in the gradient magnitude image. The algorithm tracks along these regions and suppresses any pixel that is not at the maximum (non-maximum suppression). The gradient array is now further reduced by hysteresis. Hysteresis is used to track along the remaining pixels that have not been suppressed. Hysteresis uses two thresholds: T1 and T2, with T1 > T2. If the magnitude is below T2, it is set to zero (made a non-edge). If the magnitude is above T1, it is made an edge.

And if the magnitude is between the two thresholds, then it is set to zero unless there is a path from this pixel to a pixel with a gradient above T2. This hysteresis helps to ensure that noisy edges are not broken up into multiple edge fragments.

### 3.2 *Harris corner detection based on canny edge enhancement*

First, Gaussian filter is used to perform image smoothing. Then, the sharp edge map produced by implemented canny edge detector is added to the smoothed noisy image to generate the enhanced image. Figure 1 shows the proposed block diagram to image enhancement. The Gaussian mask used in my implementation is shown below.

$$G(x, y, \sigma)=\frac{1}{2\pi\sigma^2}\exp\left(-\frac{x^2+y^2}{2\sigma^2}\right) \tag{3}$$

An edge in an image may points in a variety of directions, so we use four filters to detect horizontal, vertical and diagonal edges in the blurred image. The edge detection operator returns a value for the first derivative in the horizontal direction ($G_x$) and the vertical direction ($G_y$). From this the edge gradient and direction can be determined:

$$N=\sqrt{G_x{}^2+G_y{}^2} \tag{4}$$

$$\theta=\arctan\left(\frac{G_y}{G_x}\right) \tag{5}$$

Given estimates of the image gradients, non-maximum suppression is used to trace along the edge in the edge direction and suppress any pixel value (sets it equal to 0) that is not considered to be an edge. This will give a thin line in the output image.

Then we use Hysteresis of two thresholds (a high and a low) to eliminate streaking.

According to Eq. (1), we obtain the gray variation in each pixel $(x, y)$ of the image and compute the correlation matrix $M$ in each pixel $(x, y)$.

Where the $2\times 2$ symmetric matrix $M$ is

$$M=\begin{vmatrix} A & C \\ C & B \end{vmatrix} \tag{6}$$

where

$$\begin{cases} A=X^2\otimes G \\ B=Y^2\otimes G \\ C=(XY)\otimes G \end{cases} \tag{7}$$

Calculate two eigenvalues ($\lambda_1$, $\lambda_2$) of the matrix $M$. The two eigenvalues of the matrix $M$ determines whether the point is the corner or not.

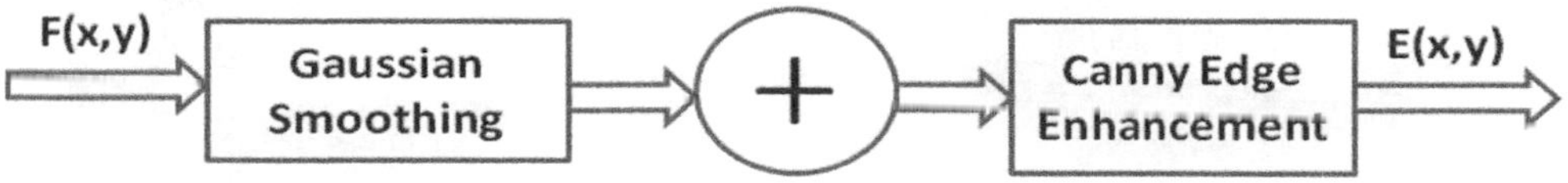


Figure 1. Block diagram of image enhancement (sharpening and de-noising) using canny edge detector.

### 3.3 *Adaptive Harris corner algorithm based on canny edge detection*

According to the analysis above, we extract feature of the corner points and its edge information based on canny edge enhancement, and the detailed algorithm process is represented as follows.

Step 1: The first step is to filter out noise in the original image use the Gaussian filter. Gaussian filter is used to perform image smoothing.
Step 2: Then, implement canny edge detector and sharp the edge of input image. The edge map produced by implemented canny edge detector is added to the smoothed noisy image to generate the enhanced image.
Step 3: Compute the correlation matrix $M$ in each pixel $(x, y)$, and then calculate two eigenvalues $(\lambda_1, \lambda_2)$ of the matrix $M$.
Step 4: Obtain the corner response function of each pixel, the response function is as follows:

$$R(x, y) = \det[M(x,y)] - k * \mathrm{trace2}[M(x, y)] \tag{8}$$

Determinant of the matrix $M$ is

$$\det[m(x, y)] = \lambda_1 * \lambda_2$$

Trace of matrix $M$ is

$$\mathrm{trace}[M(X,Y)] = \lambda_1 + \lambda_2$$

where, $k$ is an experience value, generally the value ranges from 0.04 to 0.06.
Step 5: Extract extreme points in local area by defining threshold $T$, in this paper we define $T = 0.1 * R(x, y)$ max; The corner must meet $R(x, y) > T$, sign out of the corner location with symbol "+" in the image.
Step 6: Finally, eliminate false corners in image from the detection results, display correct detection results.

## 4 EXPERIMENT AND ANALYSIS

MATLAB (Math works, MA, USA) Image processing toolbox is used in preprocessing and registering the source images. Corner detection and evaluation of the results are also performed in the MATLAB environment. The input images are captured as 3456 × 2304 bitmap image and resampled to 616 × 522-bitmap image, showing in Figure 2, (a) is the

(a) Normal image (b) Overexposed image

Figure 2. Input image.

normal image, and 2 (b) is the over-exposed one. The extracted corners are shown in Figure 3 according to classical Harris algorithm, while the results based on the new algorithm are shown in Figure 4.

As shown in Figure 3 (c) and (d), normal image can detect sufficient number of corners by Harris algorithm, corners are roughly distributing on the place where gray level changes shapely. For over-exposed image, the test result is not very ideal.

We can see that the number of detected corners increases a lot by contrasting Figure 3 (d) and 4 (f), it proves that the optimized Harris corner detection algorithm can deal with over-exposed problem. Since the joining of canny operator, the new algorithm has the image edge

(c) Normal image detection result

(d) Overexposed image detection result

Figure 3. Corner points extracted by Harris algorithms.

(e) Normal image detection result

(f) Overexposed image detection result

(g) Final result

Figure 4. Corner points extracted by the new algorithm.

Table 1. Extracted number corners points by two algorithms.

| Detection algorithm | Normal image detection result | Overexposed image detection result |
|---|---|---|
| Harris algorithm | 301 | 9 |
| Algorithm of this paper | 100 | 100 |

sharpening process, the image edge details has been enhanced, so it becomes easier to extract corners by Harris corner detection algorithm. Contrast Figure 3 (c) and 4 (e), it shows that the optimized algorithm also has good detection results for normal image, and the detection result is more accurate. Figure 4 (f) is the Harris corner detection results of over-exposed image after edge detection, the result contains a lot of false corners, so we must remove those points before the image matching. Figure 4 (g) is the final output image after getting rid of the false corners. Contrast Figure 4 (e) and 4 (g), we can see that the corner distribution on Figure 4 (g) and Figure 4 (e) are very similar, it proves that the test results is very good. We can see clearly from 3 (d), 4 (g) and Table 1 that the new algorithm in this paper has strong anti-exposure ability and less false corners.

## 5 CONCLUSION

In the actual image processing, many factors can cause input image over-exposure. In order to overcome the insufficiency of Harris corner detection method when dealing with over-exposed images. Based on the canny edge detection algorithm, this paper describes an adaptive Harris corner detection algorithm. Gaussian filter in the proposed algorithm is used to perform image smoothing, and the canny edge detector is used to produce sharp edge map. The produced sharp edge map is added to the smoothed noisy image to generate the enhanced image. Important features such as corners can be extracted from the edges of an image, therefore the new algorithm can solve the corners' information missing, location off-setting and eliminate some false corners. The experimental results show that the new algorithm has strong anti-exposure ability, meanwhile it has good detection results for normal images. The detection result can be directly applied to subsequent image matching.

## REFERENCES


Canny J.F. 1986. A computational approach to edge detection. *IEEE Transactions on Pattern Analysis and Machine Intelligence* 8(6): 679–698.

David G. Lowe. 2004. Distinctive Image Features from Scale-invariant Key points. *International Journal of Computer Vision*: 91–110.

Harris, C. & Stephens, M. 1998. A combined corner and edge detector. *Fourth ALVEY Vision Conference*: 147–151.

Kitchen, L. & Rosenfeld A. 1982. Gray level corner detection. *Pattern Recognition Letter* 3(1): 95–102.

Li Yi-bo & Li Jun-jun. 2011. Harris Corner Detection Algorithm Based on Improved Contourlet Transform. *Procedia Engineering* 15(2011): 2239–2243.

Moravec, H. 1981. Rover Visual Obstacle Avoidance. *International Joint Conference on Artificial Intelligence:* 785–790.

Mohd Azau & Mohd Azrin. 2007. Enhancement of Over-Exposed and Under-Exposed Images Using Hybrid Gamma Error Correction Sigmoid Function. *Masters Thesis*, University Putra Malaysia.

Smith, S.M. & Brady, M. 1997. A new approach to low level image processing. *International Journal of Computer Vision* 23(1): 45–78.

Stephen M. Smith,. & J. Michael Brady. 1997. SUSAN-A New approach to Low Level Image Processing. *Computer Vision* 23(1): 45–78.

*Information Systems and Computing Technology – Zhang & Gu (eds)*
*© 2013 Taylor & Francis Group, London, ISBN 978-1-138-00115-2*

# An improvement of AODV protocol based on load balancing in ad hoc networks

Xuelei Zhang & Xiaozhu Jia
*College of Information Engineering, Qingdao University, Qingdao, China*

ABSTRACT: AODV protocol is a comparatively mature on-demand routing protocol in mobile ad hoc networks. However, when the load of mobile ad hoc networks increases, the communication performance of the traditional AODV protocol sharp decline. Through the analysis and improvement of AODV protocol. This paper presents a LB-AODV protocol based on load balancing. When the packet arrives, decided to forward or discard the received packet according to the load status of each node. Finally, performance comparison of LB-AODV protocol with traditonal AODV protocol using NS-2 simulations show that LB-AODV protocol significantly increase packet delivery ratio. LB-AODV protocol has a much shorter end-to-end delay than AODV. It both optimizes the network performance and guarantees the communication quality. Thus improved the overall performance of the network.

## 1 INTRODUCTION

Mobile ad hoc network (MANET) consists of a group of mobile wireless nodes that without infrastructure support. Therefore, it can set up the network fast and relatively inexpensive. MANET has strong survivability and easy to move. MAMET make up the deficiency of cellular network and wired network. For these characteristics, it has a very broad application prospects in the field of military field, dister relief, public services, emergency search and rescue, intelligent transportation and so on (Ji Zuqin. 2010).

In the existing routing protocols, according to the different ways to create a route can divided into proactive protocol and reactive protocol. The DSDV is a typical proactive routing protocols, AODV is a typical reactive routing protocol. Studies have shown that reactive routing protocol has low routing overhead and high packet delivery ratio in the case of mobile nodes, its performance is superior than proactive routing protocols.

AODV routing protocol is designed for MANET that have dozens to thousands of mobile nodes, it perform well under light load conditions and it is able to handle the low speed, medium speed, and relatively high speed mobile rate of node in the network. However, the performance of AODV deteriorated sharply in the case of high load, a large reason is because in the choice of the path tend to use the same node as the intermediate node, large amount of data transmission through a few nodes, resulting in a high packet delay and network blocking, some nodes will be depleted of energy prematurely. For this reason, this paper proposes a improved method based on load balance to improve the overall performance of the network.

## 2 AODV ROUTING PROTOCOL

AODV is perhaps the most well-known routing protocol for MAMET and it is an on-demand distance vector routing protocol. AODV routing protocol uses reactive approach for finding routes, that is, a route is established only when it is required by any source node to transmit data packets, intermediate nodes provide the function of transmit. The protocol uses destination sequence numbers to identify the recent path. This mechanism of AODV routing protocol can be summarized route discovery and route maintenance (Wang Huabin & Luo Zhongliang. 2011).

### 2.1 *Route discovery*

When a node needs a route to send packet to another node, then it starts a route discovery process in order to estabilish a route towards the distination node. Therefore, it sends a Route REQuest message (RREQ) to its neighbouring nodes. When neighbouring nodes receive this RREQ message first determine whether or not have received the same RREQ message before, if yes, then discarded, if not, on the use of information in the RREQ message to establish a reverse route towards the source node, and then forward the message to theirs neighbours. The RREQ message will eventually reach the destination node which will react with a Route REPly message (RREP). The RREP is sent as s unicast, using the path towards the source node established by the RREQ. Therefore, at the end of the route discovery process, packets can be delivered form the source to the destination node.

### 2.2 *Route maintenance*

Node periodically broadcasts a HELLO message to determine the link connectivity information, a RREP packet with TTL = 1 is called a HELLO message. If the node can't receive HELLO message in a certain period of time, that means the link has been disconnected, then remove the routing information from the routing table which contains this node, and sends a Route ERRor message (RERR) to notify neighboring nodes and the corresponding upstream node to delete the link which is unreachable.

## 3 IMPROVED METHOD OF AODV ROUTING PROTOCOL

The analysis shows that the main overhead of AODV protocol is broadcast RREQ message when creating a route. When create a new route, you can use the route that is exists, but thest routes may being transmitted data, so it can increase the load of these routes and result in congestion. In order to overcome the above shortcomings of the AODV routing protocol, AODV is modified as follows: When an intermediate node receives a new RREQ message that TTL (Time To Live) value is greater than 0, the node treated differently according to their own load status (Gao Shengguo & Wang Hanxing. 2006).

In order to realize the above idea, use the number of packets in the buffer queue as the network load metrics (Zhu Xiaoliang & Zhen Shuli. 2008). In the process of route discovery, when the intermediate node receives a new RREQ message, according to its own load state to determine whether broadcast the received packet or discard directly. Each node in the network to monitor its buffer queue, add the value of metric to measure the load of each node. If an intermediate node buffer queue length reaches the metric value, which is no longer suitable for the transmission of new data and it is not as an intermediate node to establish a new route any more. If the buffer queue length does not reach the value of the metric, processing the received route request packet as the way of traditional AODV protocol.

Assuming that the node i of the metric value in the link is:

$$m_i = \frac{q_i}{c_i}, \tag{1}$$

where $q_i$ is the number of packets in node $i$ buffer queue, $c_i$ is the node $i$ buffer queue can accommodate a maximum number of packets.

Each node in the case of receives a RREQ message, first of all, calculate the value of $m_i$ to determine the load status of current node, and then decide whether as an intermediate node to establish a new route. Each node has three states,

i. Normal: $m_i < a$;
ii. Congestion: $a \leq m_i < b$;
iii. Paralysis: $m_i \geq b$. $a$, $b$ are coefficients of network congestion.

In order to achieve the above ideas should add a field $m_i$ in the format of routing table of the original routing protocol. Specific changes is expands the format routing table as <Destination IP Address, Destination Sequence Number, Hop Count, $m_i$, Next Hop Node, Lifetime, Routing tag>.

Each node in the network receives the packet must first calculate the current value of $m_i$ in the routing table, after that according to the value of $m_i$ and the network congestion coefficient to determine forwards or discards the received packet. If a node is in a paralyzed state, unless the current node is the destination node of this link, otherwise do not handle any routing requests and discard any received packets directly; If a node is in a congested state, only reply RREP packet when the current node is the destination node or its routing table have the information of to destination node, otherwise dropping packets in order to avoid the node into a state of paralysis; If the node is in a normal state, according to the AODV protocol processing way for processing.

Above AODV protocol improvements mainly through the addition of a simple route selection mechanism, decided to forward or discard the received packet according to the load status of each node. When establish a new route to avoid the relatively heavy load node and select relatively free load nodes and links, so that balance the network load and improve the overall performance of the whole network.

## 4 SIMULATION EXPERIMENT AND RESULT

A simulation experiment was performed by using NS2 to study the effects of the two protocols. The simulation experiment was performed on a computer with Inter Pentium G630 2.70 GHz processor and 4 GB RAM. Other parameters that were taken for simulation are shown in the Table 1 (Abdul Hadi et al. 2009).

The same experiment was repeated for the existing protocol AODV in order the compare it with the proposed protocol LB-AODV.

Information source using Constant Bit Rate (CBR), start 40 CBR streams, set metric value $a = 0.6$, $b = 0.8$. Changing the network load by changing the frequency of the information source send packet to the network.

The performance of the network was evaluated by using the following two metrics: Packet Delivery Ratio and Average End-to-End Delay (Rajender Nath & Pankaj Kumar Sehgal. 2010).

Table 1. Simulation parameter value.

| Parameter | Value | Description |
|---|---|---|
| Area size | 1200 m * 800 m | X, Y dimension in meter |
| Power range | 250 m | Node's power range in meters |
| Traffic model | CBR | Constant bit rate |
| MAC | IEEE802.11 | Medium access control protocol |
| Bandwidth | 2 MBPS | Node's bandwidth in megabits per second |
| Node placement | Uniform with 50 nodes | Node placement policy |
| Packet size | 512 Bytes | Minimum transfer unit |
| Node speed | 5 m/s | Node speed in meter per second |

### 4.1 *Packet Delivery Ratio*

Packet Delivery Ratio (PDR): It is a ratio of the data packets delivered to the destinations to those generated by the Constant Bit Rate (CBR) sources. This illustrates the level of delivered date to the destination.

$$\text{Packet Delivery Ratio} = \sum \text{Number of packet receive} / \sum \text{Number of packet send} \quad (2)$$

Testing LB-AODV and AODV at the same simulation environments, the results are shown in Figure 1.

Figure 1 shows in the case of a smaller amount of data sent to the network, the difference of packet delivery ratio of two routing protocol are not obvious. However, with the increases of network load, the packet delivery ratio of AODV protocol falling fast than LB-AODV, and simulation results also show that when the network load is heavy, the improvement effect is more obvious.

### 4.2 *Average End-to-End Delay*

Average End-to-End Delay: This includes all possible delays caused by buffering during route discovery latency, queuing at the interface queue, retransmisson delays at the MAC, and propagation and transfer times.

Testing LB-AODV and AODV at the same simulation environments, the results are presented in Figure 2.

Figure 2 shows in the case of a smaller amount of data sent to the network, two kinds of routing protocols' average end-to-end delay is very close. However, with the increases of network load, the average end-to-end delay of the two kinds of protocols are beginning to rise, but LB-AODV have a lower rate of increase of average end-to-end delay than the AODV. As can be seen from Figure 2, improved protocol can obtain lower average End-to-End Delay.

PDR and end-to-end delay are presented in Figures 1 and 2. As expected, LB-AODV significaltly echance the PDR and reduce the end-to-end delay compared with AODV. From the above analysis and statistics, with the increases of network load, improved protocol can avoid congested paths, reducing dely, improve packet delivery ratio, thus improving the overall performance of the network.

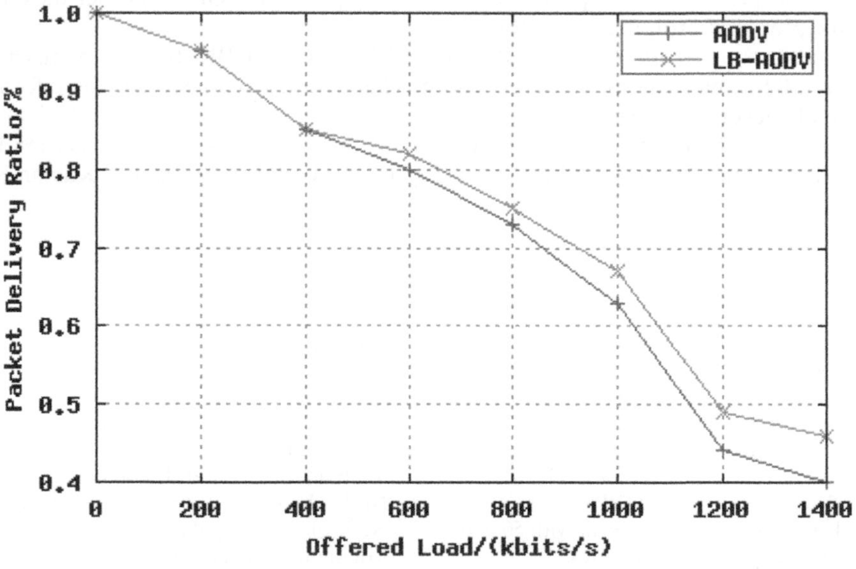


Figure 1. PDR comparison on AODV and LB-AODV.

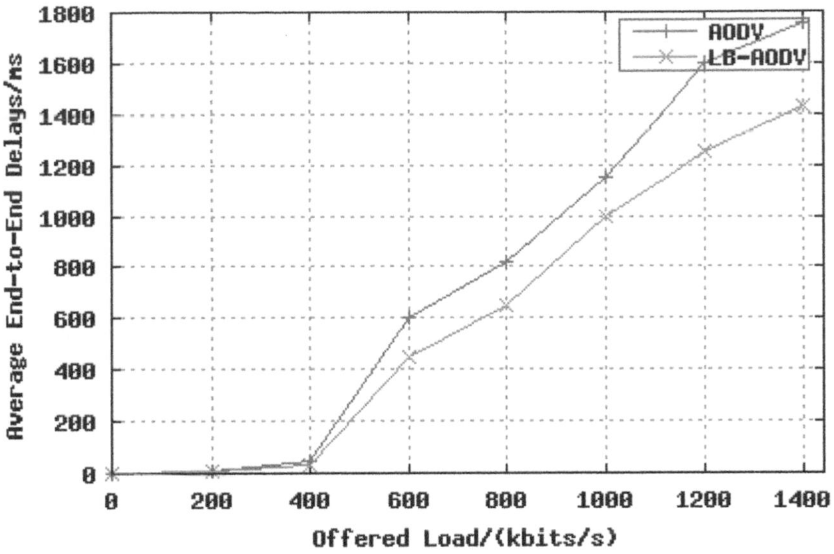


Figure 2. Average End-to-End Delay comparison on AODV and LB-AODV.

## 5 CONCLUSION

In this paper, based on AODV, proposed a new protocol LB-AODV protocol which according to the load status of the node to process the received packet. The experiment results show when the network load increases, LB-AODV has a much higher packet delivery ratio and a much shorter end-to-end delay than AODV, optimizes the network performance and also guarantees the communication quality.

In nutshell, we can say that the proposed LB-AODV protocol can avoid congested paths and reach the purposes of network load balancing. Accordingly, have a better performance than AODV in the state of heavy load.

## REFERENCES

Abdul Hadi Abd Rahman, Zuriati Ahmad Zuharnain, Performance Comparison of AODV, DSDV and I-DSDV Routing Protocols in Mobile Ad Hoc Networks, Europe Journal of Scientific Research, ISSN 1450-216X, Vol. 31, No. 4 (2009), pp. 566–576.

Gao Shengguo, Wang Hanxing, Congestion-decreased AODV Routing Protocol [J], Computer Engineering, Vol. 32, No. 23, December 2006, pp. 108–110.

Ji Zuqin, Mobile Ad Hoc Network Review [J], Science and Technology Information, No. 11, 2010, pp. 15–18.

Rajender Nath and Pankaj Kumar Sehgal, SD-AODV: A Protocol for Secure and Dynamic Data Dissemination in Mobile Ad Hoc Network, IJSSI International Journal of Computer Science Issues, Vol. 7, Issue 6, November 2010, pp. 131–138.

Wang Huabin, Luo Zhongliang. Research and Improvement of NS2-based AODV Protocol in Ad Hoc Networks [J], Computer Era, No. 5, 2011, pp. 12–14.

Zhu Xiaoliang, Zhen Shuli, Improvement of AODV routing protocol in Ad Hoc Networks [J], Computer Technology and Application Progress, 2008, pp. 1340–1343.

*Information Systems and Computing Technology – Zhang & Gu (eds)*
*© 2013 Taylor & Francis Group, London, ISBN 978-1-138-00115-2*

# Improve the vocational college students' interest in learning by optimizing the teaching contents of advanced mathematics

Hongge Zhao
*Shandong Water Polytechnic, Rizhao, Shandong, P.R. China*

ABSTRACT: "Interest is the best teacher." Many students in vocational colleges are serious shortage of confidence in learning advanced mathematics. Coupled with their lack of good study habits, these vocational students become very passive in learning advanced mathematics, and then their learning advanced mathematics are blocked. Faced with this situation, the only way to keep them out of this predicament is inspiring the students' interest in learning successfully. This paper uses examples to elaborate that can improve vocational students' interest in learning by optimizing the content of advanced mathematics teaching.

## 1 INTRODUCTION

Johann Amos Comenius pointed out that "Interest is one of the main ways to create a joyous and bright teaching and learning environment." The great scientist Albert Einstein once said: "Interest is the best teacher." This means that once a person has a strong interest in something, he would take the initiative to seek knowledge, to explore, to practice, and would produce unpleasant emotions and experience from them. This shows that how important it is to arouse students' interest of learning in teaching. Practice has proved that when students' interest in mathematics are spontaneously generated, the students will produce a strong desire for knowledge and they will be willing to learn, love to learn, and enjoy learning mathematics and they will learn better.

Recently, with the fast development of college vocational education, more and more students are rushing into colleges, flinging a great challenge to the latter in quality of the programs. The learning situation survey shows that more and more vocational students lack of interest in learning, especially in advanced mathematics. And many vocational students think that the math is too difficult, resulting in psychological fear. This situation leads to many students are serious shortage of confidence in advanced mathematics, coupled with their lack of good study habits, so that they become very passive in learning. According to this circumstance, how to simulate their interest in learning becomes very important. Based on years of teaching practice, the author found that the vocational high school students' interest in learning can be improved by optimizing the teaching content of higher mathematics.

## 2 CREATING THE "QUESTION SITUATION" TO AROUSE THE ENTHUSIASM STUDYING

According to matheaching psychology, teachers should try to make the students "hungry" for knowledge before mathematics learning, and this can stimulate students' interest, motivation and enthusiasm in learning. So in advanced mathematics teaching, a "question situation" could be created by the teachers and vocational students using the existing knowledge and problem-solving process, then the teaching points are introduceded. By using this method, the students' interest in learning can be improved greatly and make vocational students from passive to active in learning, achieve a multiplier effect.

For example, vocational students have learned how to solve the extreme function problem in senior middle school. They are able to get the extreme valve of the function which is relatively simple and derivative by the image method or derived method. However, when the function of the image is not easy to draw or the function derivative does't exist at some points, the students will not be able to obtain the extreme value of the function successfully. To solve this, teachers can create "question situation", and then targeted to carry out the teaching.

Example 1. Finding the function extreme valve: $y = f(x) = x - 3(x-1)^{2/3}$

1. Posing problem
   The teacher give a question that finding the extreme value of the function $y = f(x) = x - 3(x-1)^{2/3}$, and many students can slove it by derived method. That is, $f(9) = -3$ is the minimum of the function, and there is no maximum. Maybe, somebody could not find the results.
2. Teacher announce the answer
   The teacher announce that $f(9) = -3$ is the minimum and $f(1) = 1$ is the maximum.
3. Creat "question situation"
   The teacher ask the students why the maximum of the function is missed. (creat "question situation")
4. Explain new knowledge
   Firstly, the teacher give the general steps that how to find the extreme value of the function: ①Write the domain of definition of function. ②Find the derivative of function. ③Find the points which does not exist derivative or the derivative equal to zero in the domain of definition of function. ④List. ⑤Write conclusion.
   Secondly, regard the above problem as an example, and then teachers demonstrate the general steps of function extremum seeking. (solving process slightly)

## 3 DRIVING THE ACTIVE COGNITIVE OF VOCATIONAL STUDENTS BY "REAL-LIFE CASE"

The lecture of the advanced mathematics knowledge should be return to real life problems, and the vocational students should experience the practicality of the advanced mathematics knowledge. And then drive the active cognitive of vocational students by case to stimulate their interest in learning advanced mathematics.

For example, the continuity and nature of the function has a wide range of applications, and can solve some problems encountered in daily life.

Example 2. A chair on uneven ground, usually only three feet touch the ground, however just a little rotation angle can make four legs touch the ground at the same time, and that is stability. But how to prove this actual problem?

Analysis:

In order to describe this problem by mathematical language, we need to do some necessary assumptions on the chairs and the ground.

1. The length of the chair's four legs are same, the contact place of the chair's leg and the ground regarded as a point, the connection of the four legs is a square. (the assumption on the chair)
2. The height of the ground is continuously changed and this can be regarded as a continuous surface on the mathematical. (the assumption on the ground)
3. The ground is relatively flat, so that the chair's three legs can touch the ground at the same time. (the assumption on the relationship between the two) (process of solution slightly)

In Figure 1, in order to describe the position of chair, coordinate system is established and the center of the square is selected as the origin. Then, the chair angle of rotation

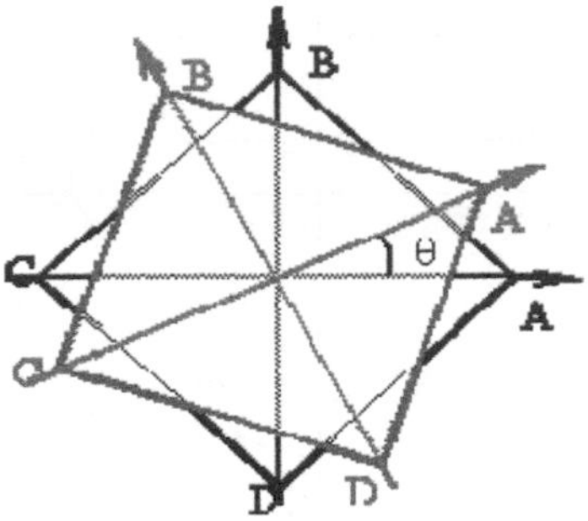


Figure 1. Coordinate system of position of chair.

can be described by the angle $\theta$. When the legs of the chair touch the ground, that is, the distance between the legs and ground is 0, which is the relationship between the chair and ground.

Let $f(\theta)$ represents the sum of distance that between legs A and C to the ground, $g(\theta)$ represents the sum of distance that between legs B and D to the ground, $h(\theta) = f(\theta) - g(\theta)$, by the nature of the continuous function that there must be a point $\theta_0(0 < \theta_0 < \pi/2)$, make $h(\theta_0) = 0$. In the square of the center to rotation of the chair, when the rotation angle is less than $\pi/2$, the chair will be able to put stability.

## 4 COMBINED WITH PROFESSIONAL TO STIMULATE THE VOCATIONAL STUDENTS' LEARNING INTEREST

After entering vocational college, the students have made it clear what is their professional subject. The learning of advanced mathematics in college should be distinguished from mathematics curriculum in senior middle school and combined with their professional. And the application of advanced mathematics in the specialized courses should be emphasized. To make the vocational students know the value of advanced mathematics career and form a correct attitude towards learning mathematics, the enthusiasm for learning could be stimulated.

For example, in the course of <Conservancy Engineering Construction>, the derivative for the extrme value is mainly used to blast crater design and layout in hydraulic engineering and building construction for hydraulic engineering professional vocational students. The so-called blasting crater (shown in Fig. 2), like a funnel, is a inverted cone blasting pit which is formed When the blasting kits have the energy that threw the part of the media into the free surface. The geometric characteristics parameters of blasting crater are as follows: the minimum resistance line $W$, the blasting action radius $R$, funnel base radius $r$, apparent depth of blasting crater $P$ and casting distance $L$ and so on. The geometric characteristics of blasting crater reflect the relationship between kits' weight and depth and the sphere of influence of the blasting.

Example 3. The blasting explosive package is commonly used in construction engineering for quarrying or borrow. Practice shows that the part of blasting is inverted conical shape as shown in Figure 3. The length of cone bus blasting radius is $R$, it is certain; The radius of the bottom surface of the cone is $r$. How deep the package of explosives should be buried when the blasting volume has the maximum?

Analysis: Firstly, according to the formula of the cone volume $V = 1/3 \cdot \pi r^2 h$, the function can be established. Then the depth of explosive package $h$ can be obtained by the method of solving extreme value of a function. (Answers, calculation slightly)

When the depth $h = \sqrt{3}/3 \cdot R$, the blasting volume has maximum.

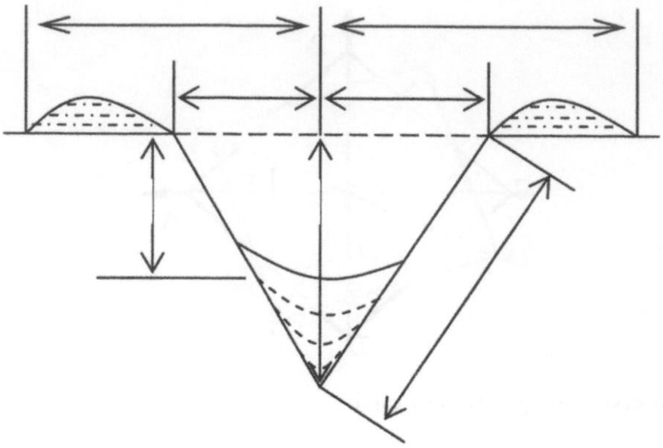

Figure 2. Blasting crater.

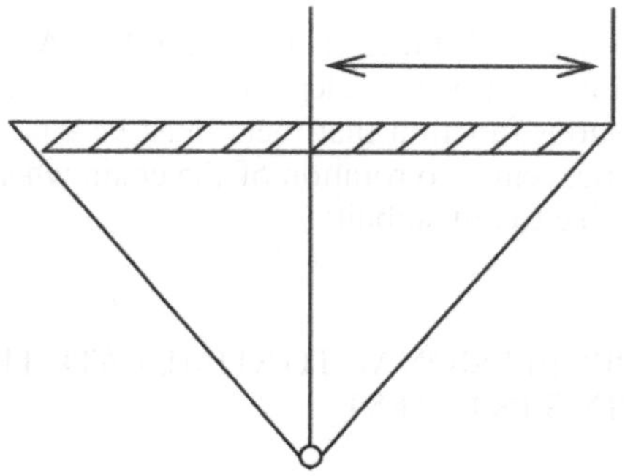

Figure 3. Explosive package.

## 5 CONCLUSION

There are many ways to stimulate the interest of vocational students in learning, not only by optimizing the content of higher mathematics teaching, but also by a variety of teaching methods. In addition, many other aspects also need to be attention, such as, to enhanced confidence of vocational students in learning advanced mathematics, to establish a good relationship between teachers and students, and to create a harmonious and pleasant atmosphere of classroom teaching.

## REFERENCES


Chuanxiang Yu. 2011. Based on the "1221" model of vocational mathematics curriculum design and Exploration. Mechanical Vocational Education. (6):35–36.

Leslie·P·Site Fu, Jerry Gail. Constructivism in education. Gao Wen, Translated. Shanghai: East China Normal University Press.

Qiang Li. 2007. Teaching thinking for function concept in new teaching material of senior middle school. Bulletin of Mathematical. 46(5):33–35.

Tao He. 2011. Services for local economic development via innovative higher vocational mathematics teaching. Chinese Vocational and Technical Education. (8):88–90.

*Information Systems and Computing Technology – Zhang & Gu (eds)*
*© 2013 Taylor & Francis Group, London, ISBN 978-1-138-00115-2*

# Research of based on the rules of the Chinese-Tibetan machine translation system dichotomy of syntactic analysis method

Zang-tai Cai
*Computer College of Qinghai Normal University, China*

ABSTRACT: Machine Translation System (MTS) is a typical nature language processing system, and language technique is a main technique in MTS. Applied MTS commonly adopts the translation measure with restrained language and based on a certain rules as a main measure. Combining with the research practice based on the 973 project—Research of Chinese-Tibetan machine translation Key technology, this paper discusses the principle which combined both word information and syntax rules. It also advances the dichotomy of syntax analysis focuses on verb. Accordingly on the range of restrained language, this paper afford a useful method to create a machine translation rule which has high adaptability and to effectively advance the efficiency of MTS' syntax analysis.

## 1 INTRODUCTION

With the growing popularity of computer technology, how to translate a large number of Chinese and English technology information, textbooks, references, and popular science readings into Tibetan in time has led the services of science and technology, education, and culture to constrain the socio-economic development in most Tibetan areas. In face of dire shortage of Chinese-Tibetan translators through technology these days, the development and promotion of the application of Chinese-Tibetan machine translation system will definitely contribute to the resolution of such problem. Language technology is the key skill in the machine translation system. The discussion of syntax analysis of machine translation system is thus an important issue.

Machine translation has many methods based on statistical machine translation, example-based machine translation and rule-based machine translation. Rule-based machine translation is the system we use for our 973 project, Chinese-Tibetan Machine Translation System. What the machine translation does is the conversion between two languages of an infinite set of sentences. Under the existing scientific level, computer science cannot theoretically prove the possibility of conversion between the source language and the target language of an infinite set of sentences by a limited rule-based machine translation system. Therefore, what is more applicable in practice is to apply restricted natural languages. The Chinese-Tibetan translation machine system we conduct research on is a kind of restricted language machine translation system.

## 2 SYSTEM STRUCTURE

Our Chinese-Tibetan machine translation system consists of three parts, technology system, document system and electronic dictionary, and uses C++ language in the Windows environment. Taking into account the needs of users in practical application, the system has pre-translation and post-translation editing function. The Chinese-Tibetan-English electronic dictionary in system has 186000 words and the translating rules of the system reach 1300.

### 2.1 *Module division*

In general, BZD Chinese-Tibetan machine translation system is classified into five parts; dictionary knowledge base maintenance, rule knowledge base maintenance, automatic word segmentation, Chinese-Tibetan translation and system settings. The main functions in the module of the figure are as follows:

- Dictionary Maintenance Module
  Responsible for managing Chinese word segmentation, Chinese-Tibetan dictionary; the main functions are the order of the dictionary, adding entry, deleting and revising. It has the function of dynamic inquiry of Chinese and Tibetan phrases as well.
- Rule Maintenance Module
  Responsible for managing rule knowledge database; the main functions are the order of rule knowledge database, adding rules, deleting and revising. It has the function of dynamic inquiry of rules according to rule No or driving words as well. In order to facilitate the management, reduce contradicting rules and improve the effectiveness of the rules, the rules are classified into two main rules (phrase rule and sentence rule), and managed and maintained at different levels.
- Segmentation and POS Tagging Module
  Apply maximum positive match method combined with segmentation rule, conduct automatic word segmentation for Chinese texts; when part of speech tagging for word segmentation results, it is divided into three kinds of labeling according to the actual corpus; label two or more than two words according to the part of speech in dictionary, label various Chinese punctuation marks, non-Chinese character strings, numerical strings (labeled as numeral type) and the classification of ASCII strings, and label the unregistered Chinese character strings (usually labeled as noun).
- Chinese-Tibetan Translation Module
  Disambiguation rules include grammatical and semantic rules. Rules of grammar disambiguation are again divided into individual and general rules. Statistical disambiguation is an auxiliary disambiguation measure of the system. The probability of different words used in different disciplines is not the same and the probabilities of what homonymous words mean in different disciplines is also not the same. The principle of "High-frequency foresight" is applied to improve relative accuracy of the system. The process of phrase combination is that the words in sentences are combined in particular order. The syntactic function and semantic information of every phrase are similar to the root word of the phrase. The system applies the method of phrase combination of isolating the root verb. The matching rule of sentence patterns is also divided into individual and general rules and the order of individual rules precedes that of general rules. The system applies simultaneously the strategies of analysis, conversion, and generation, and finally generates the Tibetan translation after a unified scan.
- System Settings Module
  System settings module provides users with the revision of system parameter and settings.

### 2.2 *Working procedures*

System working procedures as shown in Figure 1, the main functions of the parts shown in figure are as follows:

- Dictionary Database
  Dictionary applies the file format of database and the presentation of information uses multi-frame presentation. The general format of data entry structure is as follows:

  Dictionary database structure = (Chinese source sentence; controlling the information, grammatical and semantic information; Tibetan target sentence).

  Among them, Chinese source sentences include commonly used Chinese words, idioms and frequent scientific phrases. The controlling information is used to control word

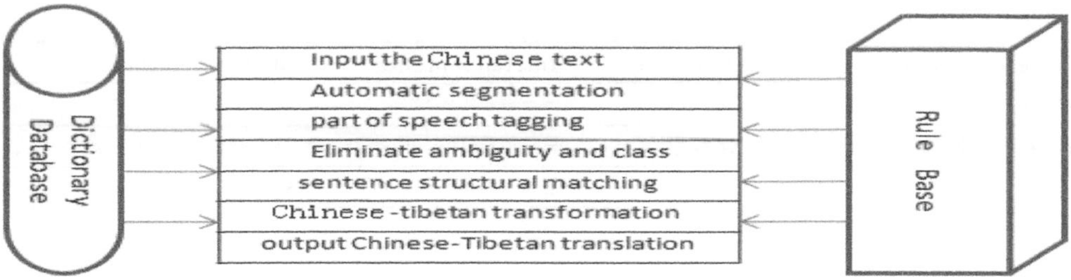


Figure 1. The working procedure of Chinese-Tibetan Machine Translation system.

processing procedure and mainly as a guiding function for processing the collocation of words and inserting dynamic phrases during the operation. Grammatical and semantic information includes the Chinese part of speech, child information, semantics, Tibetan part of speech, addition and connection of semantics and auxiliaries etc. These information, the information of the tenses of verbs in particular, will change dynamically in the process of regular combination. Tibetan translation is the Tibetan words which correspond to the Chinese original words.

- Rule Base

  The structure of data of rule is as follows;

  Rules = (<the head><the dynamic information domain><the conversion domain><the generation domain><the classification domain><the maintenance domain>).

  Among them, the head which is mainly used for Chinese analysis includes key driving words and regulate the left part and the right part. The dynamic information domain records the words combined before and after, phrase grammar, information of semantics and tenses. The conversion domain is for structural conversion between Tibetan and Chinese. The generation domain is for the addition and connection of auxiliaries. The classification domain is used to distinguish the types of rules of different properties. The maintenance domain is for rule maintenance.
- The entry of original text could be text file, keyboard input and scanning.
- Establish text segmentation and complete tagging the part of speech.
- Utilize disambiguation rules to exclude the rules of ambiguity, default the first part of speech if there is no the rules of ambiguity.
- Transfer the phrase rules of level four successively to complete phrase combination.
- Scan sentence structural rules successively to complete sentence structural matching, use signs to complete complex sentence matching.
- Scan the translation sentence by sentence, add and connect Tibetan axillarics, and output Chinese-Tibetan translation.

## 3 DICHOTOMY

As stated in introduction, language technology is the key technology in the machine translation system. The following is the discussion of some problems of syntax analysis of Chinese-Tibetan machine translation system.

### 3.1 *Combination of lexical items and grammatical rules*

With the development of technology, the position of dictionary has become more important in natural language processing system and the electronic dictionary has already become the basis of development of natural language processing application system. Dictionary as an integral part of grammar, the adjunctive information of every lexical item in the dictionary need to integrate with grammatical rules in the system application so that the driving rules

Table 1. Different word order.

| Chinese sentence | Chinese structure | Tibetan sentence | Tibetan structure |
|---|---|---|---|
| 我 是 教师 | S+V+O | ང་དགེ་རྒན་ཡིན། | S+O+V |
| 我 看了 书 | S+V+O | ང་ཡིས་དཔེ་ཆ་ལ་བལྟས། | S+auxiliary+O+auxiliary+V |
| 学生 在 教室 里 | S+V+O | སློབ་མ་སློབ་ཁང་དུ་ཡོད། | S+O+auxiliary+V |

Table 2. Different word order (Another).

| Chinese sentence | Chinese structure | Tibetan sentence | Tibetan structure |
|---|---|---|---|
| 新 兵 | adj+Center word | དམག་མི་གསར་པ། | Center word+adj |
| 三 国 | No.+Center word | རྒྱལ་ཁབ་གསུམ། | Center word+No. |
| 那个 人 | pron+Center word | མི་གན། | Center word+pron |

of lexical items could be achieved. This is because one of the core technologies of machine translation is the completion of internal structural conversion between two different languages. To complete so, it is necessary to understand the characteristics of the sentence structures of two languages. At the macro level, the comparison of sentence structures between contemporary Chinese and Tibetan has mainly three differences;

1. Different word order. Chinese order is usually SVO (S is subject, V is predicate, and O is object). Tibetan is SOV. For example, see Table 1.

   Another expression of different word order is that adjectives, numerals and pronouns when used as attributes precede the central words in Chinese language and they come after the central words in Tibetan language. For example, see Table 2.
2. Different morphological changes. Chinese language lacks morphological change whereas Tibetan language has morphological changes. The form of verb tense is an obvious difference. Tibetan verbs have tense forms. For example, the Tibetan word "eat" has three tense forms, ཟ་བ (continuous), བཟས་བ (past) 和 བཟའ་བ (future), but Chinese verbs dot not have tense forms. For example "吃" is the only form. Tense is expressed by such particular words as "了" for fast and "将" future.
3. Different ways of expression. The expression of Chinese sentences mainly depends on word order while Tibetan sentences rely on auxiliaries. Addition and connection of auxiliaries in Tibetan sentences have direct impact on the accuracy of Tibetan expressions. The following is a further comparison between Chinese verbs and Tibetan verbs.

Both Chinese and Tibetan have transitive and intransitive verbs, but there is an obvious difference. In Chinese, there is only one form of relationship between a transitive verb, subject and object, which is SVO. In Tibetan, there are many types of relationships between a transitive verb, subject and object requiring the addition of auxiliaries. In addition, Chinese transitive verbs don't have dependent and independent verbs while Tibetan transitive verbs do (the action of independent verbs can be decided subjectively whereas the action of dependent verbs cannot) and the collocation of dependent transitive verbs and intransitive verbs with subject and object has different structures. Thus, the adjunctive information of verbs in dictionary needs to reflect the characteristics of Tibetan verbs enabling combinations to abide by the grammatical rules. For example: t and I respectively represent transitive and intransitive, z, t, and 1 represent Tibetan independent transitive verbs, intransitive verbs and possessive relationship, x, 1 and k represent the relationship of three forms of Tibetan transitive verbs, subject and object, X represents agents, L represents predicate, object in Tibetan represents the target of the verb, the place of the action, the result of the action and the tools of the action as well, L also represents the auxiliaries of possessor-subject. The information

Table 3. Different ways of expression.

| Verbs | Item | C-sentence | C-structure | T-sentence | T-structure |
|---|---|---|---|---|---|
| 来到 | vttl | 我 来到 北京 | S+V+O | ང་པེ་ཅིན་དུ་སླེབས། | S+O+L+V |
| 保持 | vttx | 我们 保持 联系 | S+V+O | ང་ཚོས་འབྲེལ་བ་རྒྱུན་འཁྱོངས་བྱེད། | S+X+O+V |
| 研究 | vttk | 我 研究 问题 | S+V+O | ངས་གནད་དོན་ལ་ཞིབ་འཇུག་བྱེད། | S+X+O+L+V |
| 有 | vtl | 我 有 书 | S+V+O | ང་ལ་དཔེ་ཆ་ཡོད། | S+L+O+V |
| 来自 | vtz | 我 来自 青海 | S+V+O | ང་མཚོ་སྔོན་ནས་འོང་། | S+O+L+V |
| 得出 | vitl | 我 得出 结论 | S+V | ང་ལ་སྒོམ་ཚོག་ཐོབ། | S+L+V |
| 讲课 | vitx | 老师 讲课 | S+V | དགེ་རྒན་གྱིས་སློབ་ཁྲིད་བྱེད། | S+X+V |

of lexical items of part-of-speech tagging of verbs could be related to the information of Tibetan grammatical structure as shown in Table 3.

As the above example has shown, the combination of the information of lexical items and grammatical rules allows a rule-based system of a larger adaptive machine translation within a range of restricted language.

### 3.2 *Syntactic analysis dichotomy*

Any machine translation system is only possible after a long-term testing and improvement. This is because an addition of a commonly used word is likely to add new rules or modify the original rules. Consequently, new rules often lead to recombination and readjustment of the rules of the whole system. Therefore, minimizing the adjustment workload or shortening the adjustment time is an important issue in the development of machine translation system. The combination of BZD Chinese-Tibetan document machine translation system introduces a form of generating method of sentence analysis. It can effectively reduce the number of rules so that the time of reorganization of rules and adjustment could be reduced.

Existing machine translation systems use a sentence as a unit to do the translation and grammatical analysis algorithm is also a result of the analysis of sentences. Since both Chinese and Tibetan have subject, predicate and object, trichotomy is applied and the sentences are combined and converted according to subject, predicate and object. We apply syntax analysis dichotomy during the development of BZD Chinese-Tibetan machine translation system and it is as effective as trichotomy. But the number of grammar rules is greatly reduced. There are two following points to explain it:

1. Basic Way of Syntax Analysis Dichotomy
   As we can see from the discussion of combination between the information of lexical items and grammatical rules, the basic structure of Chinese sentences is SVO and it is basically a structure of word order. But Tibetan is different. Although the basic structure is SOV, the structure has many forms, for example; SOLV, SXOV, SXOLV, SOV, SLOV etc. The generations of predicate-object and subject-predicate phrases facilitates the general collocation of subject, predicate and object in Tibetan language. It is also applicable to Chinese language even if the synthesis of SVO is completed through synthesis of predicate-object and subject-predicate phrases. Two points need to be highlighted when synthesizing: The first point is verb is based on the understanding of the pivot and the axis of sentence structure setting the integrated word of verb-object and subject-predicate phrases as verb. The second point is the generation of verb-object phrases should be prioritized over that of subject-predicate phrases. By so doing, SVO in a sentence can be presented by subject-predicate phrase. This is because a complete sentence with only subject and predicate is enough. For example; the Chinese structure of "**我是学生**" is SVO.

Table 4. Different ways of expression.

| C-sentence | C-sentence structure | T-sentence | T-sentence structure |
|---|---|---|---|
| 我学习物理学 | S+V+O | ངས་དངོས་ལུགས་ལ་སློབ་སྦྱོང་བྱེད། | S+X+O+L+V |
| 我学习 | S+V | ངས་སློབ་སྦྱོང་བྱེད། | S+X+V |
| 学习物理学 | V+O | དངོས་ལུགས་ལ་སློབ་སྦྱོང་བྱེད | O+L+V |
| 要学习物理学 | U+V+O | དངོས་ལུགས་ལ་སློབ་སྦྱོང་བྱེད་དགོས། | O+L+V+U |
| 一定要学习物理学 | D+U+V+O | ངེས་པར་དུ་དངོས་ལུགས་ལ་སློབ་སྦྱོང་བྱེད་དགོས། | D+O+L+V+U |
| 我要学习物理学 | S+U+V+O | ངས་དངོས་ལུགས་ལ་སློབ་སྦྱོང་བྱེད་དགོས། | S+X+O+L+V+U |
| 我一定要学习物理学 | S+D+U+V+O | ངས་ངེས་པར་དུ་དངོས་ལུགས་ལ་སློབ་སྦྱོང་བྱེད་དགོས། | S+X+D+O+L+V+U |
| 要学习 | U+V | སློབ་སྦྱོང་བྱེད་དགོས། | V+U |
| 我要学习 | S+U+V | ངས་སློབ་སྦྱོང་བྱེད་དགོས། | S+X+V+U |
| 我一定要学习 | S+D+U+V | ངས་ངེས་པར་དུ་སློབ་སྦྱོང་བྱེད་དགོས། | S+X+D+V+U |

No matter it is SVO—S+VO, or V+O—V, SV includes SVO. In the case of Tibetan, the sentence structure is SOV. No matter it is SOV—S+OV, or O+V—V, SV includes SOV as well. The Chinese structure of "我有书" is SVO which can be presented by verb-object phrase V(有)+O(书)—V(有) and subject-predicate phrase S(我) + V(有). In the case of Tibetan, the structure of the sentence is SLOV which can be presented by verb-object phrase O+V—V and subject-predicate phrase SLV: SLOV—S+L+V—S+L+O+V.

2. The advantages of applying syntax analysis dichotomy
As stated above, the addition of new rules often lead to recombination and readjustment of the rules of the whole system. Therefore, reducing the number of new rules is in favor of the adjustment of machine translation system and using syntax analysis dichotomy with the verb as the center also contributes to the general reduction of the rules. As shown by the following example, trichotomy uses one grammatical rule for "我学习物理学" whereas dichotomy uses two grammatical rules. But the rules that dichotomy need for the rest of sentences are greatly reduced. The lexical item information for the word "学习" is vvttk. Therefore the structure of S+X+O+L+V is applied into Tibetan SOV relationship.

The sentences above usually need 10 grammatical rules. After applying syntax analysis dichotomy, V+O, U+V, D+V and S+V rules alone can achieve the function of ten generating and converting rules.

We know that scientific language is more standard and the grammatical phenomenon for document corpus is more sophisticated. Chinese-Tibetan machine translation system applies trichotomy and document translation system uses dichotomy. The testing results show that the sustainability of scientific translation system and that of document translation system are almost the same for a non-enclosed corpus. But document translation system has relatively less grammatical rules than scientific translation system showing the advantage of dichotomy.

## 4 CONCLUSION

The application of machine translation is not only very promising in the field of language information processing, but a challenging research topic as well. This article introduces the structure of Chinese-Tibetan machine translation system. After being tested and tried by some experts and users, it is shown that the system has already reached the level of practicality. The analysis of verbs is the key to sentence structure in Chinese-Tibetan machine translation system. With verb-predicate as the axis, the integration of grammatical analysis and semantic analysis can be achieved through the combination of lexical item information with grammatical rules and the integration of grammatical analysis and semantic analysis, and the application of syntax analysis dichotomy can improve greatly the efficiency of grammatical analysis by machine translation.

## REFERENCES


Dougai Cailang, Li Ynagfu. 1997. "Brief Analysis on Chinese-Tibetan Scientific Machine Translation System," Progress in the Interface and Application of Intelligent Computer. Electronic Industry Press.

Dougai Cailiang, 1995. "Brief Analysis on the Classification of Tibetan Terms with the Focus on Phrases in Chinese-Tibetan Machine Translation System." Progress in the Interface and Application of Intelligent Computer. Qinghua University Press.

Ma zhongwei. 1994. The New Theory of Natural Language Machine Translation. Yuwen Press.

Yu shiwen. 1998. The Detailed Information of Contemporary Chinese Grammar Information. Qinghau University Press.

*Information Systems and Computing Technology – Zhang & Gu (eds)*
*© 2013 Taylor & Francis Group, London, ISBN 978-1-138-00115-2*

# Query optimum scheduling strategy on network security monitoring data streams

Ying Ren
*Naval Aeronautical and Astronautical University, Yantai, China*

Hua-wei Li
*Naval Aeronautical and Astronautical University, Yantai, China*
*Shandong Business Institute, Shandong, Yantai, China*

Hong Lv & Hai-yan Lv
*Naval Aeronautical and Astronautical University, Yantai, China*

ABSTRACT: This thesis studies on the query mechanism of network security monitoring data streams and proposed a method of using flow data as materialized sharing intermediate results of compression, storage flow data in memory. Optimizing the storage efficiency of the system in order to save intermediate results more. Through the experimental verification of the data compression of multiple query optimization is effective.

## 1 INTRODUCTION

In recent years, the application fields of information processing technology has been widely expanded, and a variety of network security threats are more and more, and the network security for the purpose of monitoring and analysis of monitoring data, has the important practical significance. Network security monitoring produced massive, sustained, rapid data. The real-time data can be sustained, rapid analysis using stream processing technology, and the technology of data stream processing has become a hotspot in the field of database research (Chien JT, Wu CC, 2002).

When large-scale network security monitoring data stream processing system in the management and analysis of security monitoring, multiple system will also landing system, and produce a large number of queries, and each query processing so that this kind of query is low (Zhao Li, Zheng Wen-ming, Zou Cai-rong, 2004). Multiple queries may share the same subtask and through the optimization can improve the query efficiency, reduced computational overhead multiple query. Multiple query optimization generally use two methods: optimal query plans and materialized sharing intermediate results. This paper deeply research in data stream query processing mechanism, and put forward a kind of method to stream data as materialized sharing intermediate results of compression, storage flow data in memory, optimize the storage efficiency of the system in order to save intermediate results more.

## 2 MULTIPLE QUERY OPTIMIZATION IN DATA STREAM

Given a query set $Q = \{Q_1, \ldots, Q_q\}$ consists of complex SQL query $Q_1, \ldots, Q_q$, and contains the relationship set $R_1, \ldots, R_r$, each query $Q_i$ is a subset of the relation of attribute sets. $|R_l|$ represents the group number. In relations $R_i$. Data stream query processing engine only allows sequential, one-pass data stream list $R_1, \ldots, R_r$, and cannot be read again. Multiple query optimization technique using materialized sharing intermediate results, the subexpression computer results in multiple query the shared cache in memory for multiple query repeatedly read, suitable for dataflow characteristics of single pass data, at the same time, improve the query efficiency. Materialical shared intermediate results of query optimization technology system architecture is shown in Figure 1.

Data flow DS1, ..., DSr (also $R_1, ..., R_r$) continue to flow into the data stream query engine, and multiple queries $Q_1, ..., Q_q$ registrate to search engine, and optimize the evaluation of the query, and execuet query plan (Prasan Roy, S. Seshadri, S. Sudarshan, 2000). The query engine will be materialized intermediate results in the cache memory, a plurality of query access materialized intermediate results will get their results.

Compressed materialized intermediate results are stored in the memory of repeated use. Along with the query time window increases, storage space will increase. The limited memory space cannot meet the storage requirements increasing, while compression method of sharing intermediate results can be used to solve this problem. Materialized intermediate results can be mapped to a portion of the flow data when we define all of the query as aggregation query. Better flow data compression method should satisfy three conditions:

1. The data removal of Fang Rongyu compression.
2. Query does not need to decompress.
3. Targeted precomputed queries on data streams.

## 3 STREAM DATA COMPRESSION STRUCTURE

Select the proper flow data compression structure can save more information in the same memory. There are various compression structure such as QC-tree (Jian Pei, Laks V.S. Lakshmanan, Yan Zhan, 2003), Dwarf (Antonios Deligiannakis, Nick Roussopoulos, Yannis Sismanis, 2002) etc, and these compression structure is completely to the static data of data. Stream data of dynamic data is stored in the memory, and dynamic data is continuously generated, infinite. Compression structure cannot accommodate the entire fully materialized data compression structure in the limited memory space, and can only be stored within a certain period of time the flow data section. In order to preserve the flow data slice more in memory, wo need to further reduce the capacity of single slice.

In this paper, flow analysis requirements proposed implementation framework of data stream flow data system for real-time data, and realize the compression flow data structure of StreamQCTree (Gan Liang, Han Wei-Hong, Jia Yan, Liu Dong-Hou, 2011), and its update and query algorithm.

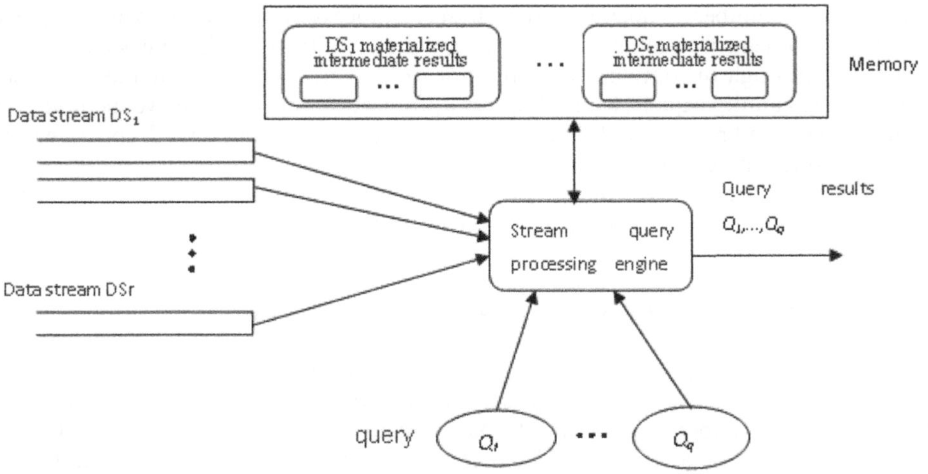


Figure 1. Multiple query optimization of data flow.

*Definition 1: StreamQCTree* StreamQCTree is similar to QC-Tree, except for the following properties:

1. The root node directly connects each time node with the edge in StreamQCTree. Each time node includes a QC-Tree tree, the sub tree corresponding data time slice $SC_{ti}$.
2. StreamQCTree contains all the basic upper class, and select add some additional last class.
3. Each node in the StreamQCTree contains an additional value cos *t*, represent the cost of search nodes. For material node, the cost is 1; For non-material node, when represented in the aggregation query, access the time of the subsequent node.
4. All the non-material nodes do not store its measurement.

## 4 MATERIALIZED NODE SELECTION

*Definition 2 Materialized node selection problem* (Anand Rajaraman, Jeffey D, Venky Harinarayan, 1996) Given a database pattern *R*, constraint condition *T*, query set *Q* and cost evaluation function C, selection of materialized node is to choose a materialized node sets *V* above the *R* whose cost *C*(*R*, *V*, *Q*) is minimization in the case of meeting the conditions of *T*.

### 4.1 *Cost calculation*

A $|D|$ dimension data, data storage allocation *m* storage unit, data time window *W*, data stream sliding window *s*. Each data time slice distribution of memory space $M_{sc} = m * s / W$. A basic upper class accounted $M_{bas}$ for a total of memory. The additional upper class can use the space for the: $M_{add} = M_{sc} - M_{bas}$.

A pointer is stored in *P*, dimension marking memory for *La*, metric value is stored for *Me*, its subsequent nodes for cos *t*. If adding an additional upper class, will increase the storage space for the

$$M_{aud} = \cos t \times P + Me + x \times La \tag{1}$$

where *x* denotes the upper dimension standard notation, determine the value for the general dimension marking sequence of the upper bound of the first * dimension mark number, $0 \leq x < |D|$.

### 4.2 *The calculation of earnings*

For any additional upper bound $UB_i$, if the access $UB_i$ in the tree nodes probability $P_{ubi}$ in the query, has materialized to access number $C_m(UB_i)$, the number of access for non-material $UB_i$ is $C_{um}(UB_i)$. The earnings rate of Material bound $UB_i$ is $B_{ubi}$.

$$B_{ubi} = P_{ubi} \times (C_{um}(UB_i) - C_m(UB_i)) / M_{aud} \tag{2}$$

$B_{ubi}$ is a correlative value which can be evaluated material payoffs with $UB_i$ respect to other upper bound so that to evaluate and select additional upper class of the best, and to achieve the minimum average query response time.

## 5 UPDATE AND QUERY COMPRESSED DATA STREAM PARTY

### 5.1 *Stream QCTree update algorithm*

StreamQCTree uses the time slice data model. In the update algorithm, data flow continuously arrived because each time interval $\Delta t$ will generate a the data time slice QC-Treesubtree. Update requires to add the subtree of the Ti time to the Root node of StreamQCTree.

Remove only has a moment. When the time slice of $T_i$ data in memory is expired, the subtree is removed from memory. As long as the QC-Tree subtree memory release and remove the pointer Root points to a subtree.

### 5.2 *StreamQCTree query algorithm*

StreamQCTree query algorithm is similar to the QC-Tree, because some nodes have not materialized, when the query which is involved non-materialized Node will generate queryanomaly. After the query abnormalities in positioning, to determine the basic upper class node which need to access in layer $i-1$ according to the hierarchy of i abnormalities in the tree, namely the function traverBUB().

Function traverBUB(*sqt, i, s[ ]*)
  For each BUB label of layer *i* in the *sqt* tree
    For from the $i+1$ layer to |D| layer in the tree
      traver(*sqt*, *s*[$i+1$]); //Look at the $i+1$ layer
      if *i*==|D|
        *result'* = getresult();
    *result* = fun(*result'*); //Function fun including sum(), count (), AVG ()
  Return *result.*

## 6 EXPERIMENT

The algorithm which is used in the experiments is VC++6.0. The operating environment is to install the Windows XP Professional PC and hardware configuration of 160 G harddisk, 2 G memory. Assume that tuples in data stream is uniformly distributed in the time dimension. The data consists of 6 dimensions, and each dimension of the data base was 100, and time window $W = 10$, Zipf (factor = 1.2), and each time a new group to reach a total of 100 k, and allocate 200 M memory space, each time allocated memory 10 M. The experiment adopted two kinds of flow data storage structure of StreamQCTree-D and QC-Tree, StreamQCTree-D with dynamic selection scheme of StreamQCTree.

The experimental results verify the effectiveness of StreamQCTree compression method, as shown in Figure 2. The StreamQCTree relative to the QC-Tree furtherly reduce the memory footprint. StreamQCTree-D choices part of materialized nodes using a dynamic choice program which materializes the nodes of the highest frequency of the query tree, trying to cut the tree nodes whose query frequency is low or which are not queried. The query response time gap is smaller relative to QC-Tree, and to reduce the performance of the query, as shown in Figure 3.

Therefore, the dynamic scheme of StreamQCTree relative to QC-Tree in the lower part of the query performance to get higher data compression rate.

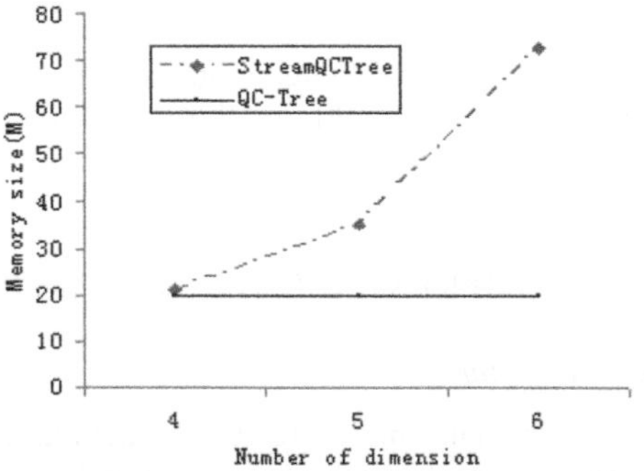


Figure 2. Comparison of memory space.

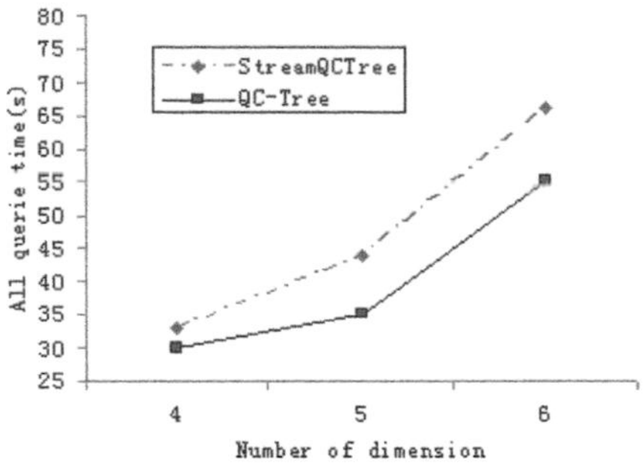


Figure 3. Comparison of query response time.

# REFERENCES


Anand Rajaraman, Jeffey D, Venky Harinarayan. Ullman. Implementing Data Cubes Efficiently [C]. Proc of the 1996 ACM SIGMOD International Conference on Management of Data, 1996, Montreal, Canada: ACM:205–216.

Antonios Deligiannakis, Nick Roussopoulos, Yannis Sismanis, et al. Dwarf: Shrinking the petacube [C]. Proc of the 2002 ACM-SIGMOD international conference management of data, 2002, madison, Wisconsin: ACM:464–475.

Chien JT, Wu CC. Discriminant Wavelet faces and Nearest Feature Classifiers for Face Recognition [J]. IEEE Trans. on PAMI, 2002, 24(12):1644–1649.

Gan Liang, Han Wei-Hong, Jia Yan, Liu Dong-Hou. StreamQCTree: A stream data compression structure [J]. Computer engineering and Applications, 2011.07.

Jian Pei, Laks V.S. Lakshmanan, Yan Zhan. QC-trees: An efficient summary structure for semantic OLAP [C]. Proc of the 2003 ACM SIGMOD International Conference on Management of Data, 2003, California: ACM:64–75.

Prasan Roy, S. Seshadri, S. Sudarshan, et al. Efficient and Extensible Algorithms for Multi Query Optimization [C]. Proceedings of the 19th ACM SIGMOD International Conference on Management of Data, 2000, Dallas, USA: ACM Press:249–260.

Zhao Li, Zheng Wen-ming, Zou Cai-rong. Face Recognition Using Two Novel Nearest Neighbor Classifiers [C]. Proceedings of ICASSP. 2004:725–728.

*Information Systems and Computing Technology – Zhang & Gu (eds)*
*© 2013 Taylor & Francis Group, London, ISBN 978-1-138-00115-2*

# Arithmetic labeling about the Graph $C_{4k,i,n}$

Ergen Liu, Qin Zhou & Wei Yu
*School of Basic Sciences, East China Jiaotong University, Nanchang, Jiangxi, P.R. China*

ABSTRACT: For a $(p, q)$ graph $G$, if there is a mapping $f$ (called the vertex label) from $V(G)$ to the set of nonnegative integer $N_0$, meet: (1) $f(u) \neq f(v)$, which $u \neq v$ and $u,v \in V(G)$, (2) $\{f(u)+f(v) \mid uv \in E(G)\} = \{k, k+d, \ldots, k+(q-1)d\}$, then said graph $G$ is called the $(k,d)$—arithmetic graph. The graph composed with several rings is a kind of important and interesting graph, in this paper, we obtain the arithmetic labeling of graph $C_{4k,i,n}$, thus we prove that the graph $C_{4k,i,n}$ is $(d,2d)$—arithmetic graph and obtain the graph $C_{4k,i,n}$ is odd arithmetic graph. At last, in order to explain the correctness of the aforementioned labels, we give the arithmetic labeling of the graph $C_{12,5,3}$ and $C_{16,6,3}$.

## 1 INTRODUCTION

The graph composed with several rings is a kind of important and interesting graph, many scholars studied on the gracefulness of this kind of the graph. In this paper, we researched the arithmetic labels of the kind graphs. The graph of $n$ kinds $C_{4k}$ with one common point was given the arithmetic labeling (Ergen, L. & Dan, W. 2009), and the graph of $C_{8,i,n}$ $(i=1,2,3,4)$ was $(d,2d)$—arithmetic graph, Ergen, L. & Dan, W. 2010). In this paper, on the basis of above arithmetic labeling to further expand, we researched the arithmetic labels of $C_{4k,i,n}$. The graph in this paper discussed is undirected, no multiple edges and simple graph. Let $G$ be a graph with vertex set $V(G)$ and edge set $E(G)$.

*Definition 1*: For a $(p, q)$ graph $G$, if there is a mapping $f$ (called the labeling of vertex) from $V(G)$ to the set of nonnegative integer $N_0$, meet:

1. $f(u) \neq f(v)$, which $u \neq v$ and $u, v \in V(G)$;
2. $\{f(u)+f(v) \mid uv \in E(G)\} = \{k, k+d, \ldots, k+(q-1)d\}$.

Then we call graph $G$ is $(k,d)$—arithmetic graph.

*Definition 2*: For definition 1, when $k=1$, $d=1$ and $f(u)+f(v) \in \{1,2,\ldots,q\}$, then we call the graph $G$ is strongly harmonious graph.

*Definition 3*: For definition 1, when $k=1$, $d=2$ and $f(u)+f(v) \in \{1,3,\ldots,2q-1\}$, then we call the graph $G$ is odd arithmetic graph.

*Definition 4*: We called that the graph composed by $n$ kinds $C_{4k}$ with $i$ common point is $C_{4k,i,n}$.

## 2 MAIN RESULTS AND CERTIFICATION

*Theorem 1*: When $1 < m \leq k$, the graph $C_{4k,2m-1,n}$ is $(d,2d)$—arithmetic graph.

Proof: As the graph shown on Figure 1.

Label all vertices as follows.

1. If $i=0,2,4,6,\ldots$

When $1 \leq j \leq 2k$

$$f(v_{i,j}) = [8k-4(m-1)]\frac{i}{2}d + (j-1)d.$$

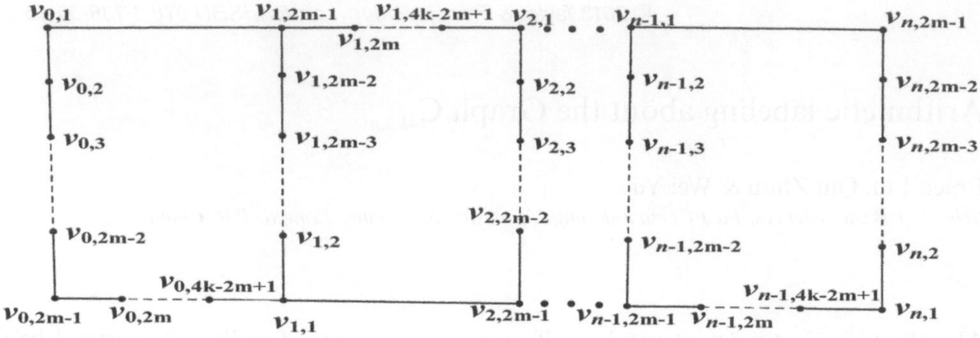


Figure 1. Graph $C_{4k,2m-1,n}$.

When $2k< j\le 4k-2m$

$$f(v_{i,j})=\begin{cases}[8k-4(m-1)]\frac{i}{2}d+(j+1)d, & j=1,3,5,\ \dots,\ 4k-2m+1\\ [8k-4(m-1)]\frac{i}{2}d+(j-1)d, & j=2,4,6,\dots,\ 4k-2m\ .\end{cases}$$

2. If $i=1,3,5,7,\ \dots$

$$f(v_{i,j})=\begin{cases}[8k-4(m-1)]\frac{i-1}{2}d+(4k-2m+j)d, & j=1,3,5,\ \dots,\ 4k-2m+1\\ [8k-4(m-1)]\frac{i-1}{2}d+(4k-2m+2+j)d, & j=2,4,6,\ \dots,\ 4k-2m.\end{cases}$$

We can see the mapping $f$ meet $f(u)\ne f(v)$ which $u\ne v$ and $u,v\in V(C_{4k,2m-1,n})$.
Next we prove that $\left\{ f(u)+f(v)\,|\,uv\in E\left(C_{4k,2m-1,n}\right)\right\}$ is an arithmetic labeling in the way of mathematic induction.
When $n=1$

Then $f(v_{0,1})=0, f(v_{0,2})=d, f(v_{0,3})=2d, \dots, f(v_{0,2k})=(2k-1)d, f(v_{0,2k+1})=(2k+2)d,$
$f(v_{0,2k+2})=(2k+1)d, \dots, f(v_{0,4k-2m-1})=(4k-2m)d,$
$f(v_{0,4k-2m})=(4k-2m-1)d, f(v_{0,4k-2m+1})=(4k-2m+2)d, f(v_{1,1})=(4k-2m+1)d,$
$f(v_{1,2})=(4k-2m+4)d, \dots,\ f(v_{1,2m-2})=4kd,\ f(v_{1,2m-1})=(4k-1)d.$

Therefore $\left\{f(u)+f(v)\,|\,uv\in E\left(C_{4k,2m-1,1}\right)\right\}$

$=\{f(v_{0,1})+f(v_{0,2}), f(v_{0,2})+f(v_{0,3}), \dots, f(v_{0,2k})+f(v_{0,2k+1}),$
$f(v_{0,2k+1})+f(v_{0,2k+2}), \dots,\ f(v_{0,4k-2m})+f(v_{0,4k-2m+1}), f(v_{0,4k-2m+1})+f(v_{1,1}),$
$f(v_{1,1})+f(v_{1,2}), \dots, f(v_{1,2m-3})+f(v_{1,2m-2}), f(v_{1,2m-2})+f(v_{1,2m-1})\}$
$=\{d,\ d+1\times 2d,\ d+2\times 2d, \dots,\ d+2k\times 2d,\ d+(2k+1)\times 2d, \dots,\ d+(2k-2m-1)\times 2d,$
$d+(2k-2m)\times 2d,\ d+(2k-2m+1)\times 2d, \dots,\ d+(4k-2)\times 2d,\ d+(4k-1)\times 2d\}$

is an arithmetic progression and the common difference is $2d$.

1. Suppose when $n = r-1$ and $r$ is even number, we know

$$\left\{f(u)+f(v)\,|\,uv \in E\left(C_{4k,2m-1,r-1}\right)\right\}$$
$$=\{d,\ d+1\times 2d,\ d+2\times 2d, \ldots,\ d+[(4k-2m+2)(r-2)+4k-1]\times 2d\}$$

is an arithmetic progression and the common difference is $2d$.

Then when $n = r$

$$\left\{f(u)+f(v)\,|\,uv \in E\left(C_{4k,2m-1,r}\right)\right\}$$
$$=\left\{f(u)+f(v)\,|\,uv \in E\left(C_{4k,2k-1,r-1}\right)\right\} \cup \{f(v_{r-1,2m-1})+f(v_{r-1,2m}), \ldots,$$
$$f(v_{r-1,4k-2m+1})+f(v_{r,1}), f(v_{r,1})+f(v_{r,2}), f(v_{r,2})+f(v_{r,3}), \ldots,$$
$$f(v_{r,2m-3})+f(v_{r,2m-2}), f(v_{r,2m-2})+f(v_{r,2m-1})\}$$
$$=\{d,\ d+1\times 2d,\ d+2\times 2d, \ldots, d+[(4k-2m+2)(r-2)+4k-1]\times 2d\} \cup$$
$$\{d+[(4k-2m+2)(r-2)+4k]\times 2d,\ d+\left[(4k-2m+2)(r-2)+4k+1\right]\times 2d, \ldots,$$
$$d+\left[(4k-2m+2)(r-2)+8k-2m+1\right]\times 2d\}$$
$$=\{d, d+1\times 2d,\ d+2\times 2d, \ldots, d+\left[(4k-2m+2)(r-2)+8k-2m+1\right]\times 2d\}$$

is an arithmetic progression and the common difference is $2d$.

2. When $r$ is odd number, By (1) we know that $\left\{f(u)+f(v)\,|\,uv \in E\left(C_{4k,2m-1,r}\right)\right\}$ is an arithmetic progression and the common difference is $2d$.

To sum up, for the arbitrary $n \in N_0$, the mapping $f: V(C_{4k,2m-1,n}) \rightarrow N_0$ meets:

1. $f(u) \neq f(v)$, which $u \neq v$, and $u, v \in V(G)$;
2. $\left\{f(u)+f(v)\,|\,uv \in E(G)\right\}$
$$=\left\{d, d+1\times 2d, d+2\times 2d, \ldots,\ d+[4kn-(n-1)(2m-2)-1]\times 2d\right\}.$$

Therefore when $1 < m \leq k$, the graph $C_{4k,2m-1,n}$ is $(d,2d)$—arithmetic graph.

*Theorem 2*: when $1 \leq m \leq k$, The graph $C_{4k,2m,n}$ is $(d,2d)$—arithmetic graph.

Proof: As the graph shown on Figure 2.

Label all vertices as follows.

1. If $i = 0,2,4,6,\ldots$

When $1 \leq j \leq 2k$

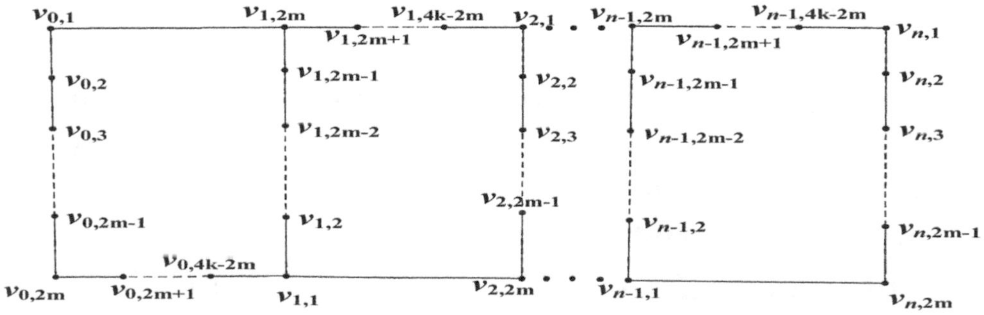


Figure 2. Graph $C_{4k,2m,n}$.

$$f(v_{i,j})=(8k-4m+2)\frac{i}{2}d+(j-1)d.$$

When $2k< j\le 4k-2m$

$$f(v_{i,j})=\begin{cases}(8k-4m+2)\frac{i}{2}d+(j-1)d, & j=2,4,6,\ \dots,\ 4k-2m\\(8k-4m+2)\frac{i}{2}d+(j+1)d, & j=1,3,5,\ \dots,\ 4k-2m-1.\end{cases}$$

2. If $i=1,3,5,7,\ \dots$

When $1\le j\le 2k+3$

$$f(v_{i,j})=\begin{cases}(8k-4m+2)\frac{i-1}{2}d+(4k-2m-1+j)d, & j=2,4,6,\ \dots,\ 4k-2m\\(8k-4m+2)\frac{i-1}{2}d+(4k-2m+1+j)d, & j=1,3,5,\ \dots,\ 4k-2m-1.\end{cases}$$

When $2k+3< j\le 4k-2m$

$$f(v_{i,j})=(8k-4m+2)]\frac{i}{2}d+(4k-2m+1+j)d.$$

We can see the mapping *f* meet $f(u)\neq f(v)$ which $u\neq v$ and $u,v\in V(C_{4k,2m,n})$.

By theorem 1, we know that $\{f(u)+f(v)\,|\,uv\in E(C_{4k,2m,n})\}$ is an arithmetic progression and the common difference is $2d$.

Therefore when $1\le m\le k$, the graph $C_{4k,2m,n}$ is $(d,2d)$—arithmetic graph.

*Inference 1*: $C_{4k,i,n}$ is an odd arithmetic graph.

Proof: Due to the graph $C_{4k,i,n}$ is $(d,2d)$—arithmetic graph, therefore when $d=1$, then $f(u)+f(v)\in\{1,3,\dots,2q-1\}$, By theorem 1, we know that $C_{4k,i,n}$ is an odd arithmetic graph.

## 3 LABELING OF SOME SPECIAL GRAPH

In order to explain the correctness of the aforementioned labels, we give the arithmetic labeling of the graph $C_{12,5,3}$ and $C_{16,6,3}$.

1. The arithmetic labeling of the graph $C_{12,5,3}$ on Figure 3.
2. The arithmetic labeling of the graph $C_{16,6,3}$ on Figure 4.

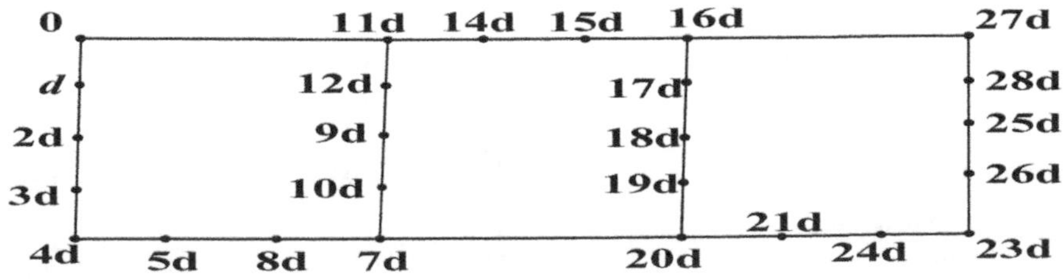


Figure 3. Graph $C_{12,5,3}$.

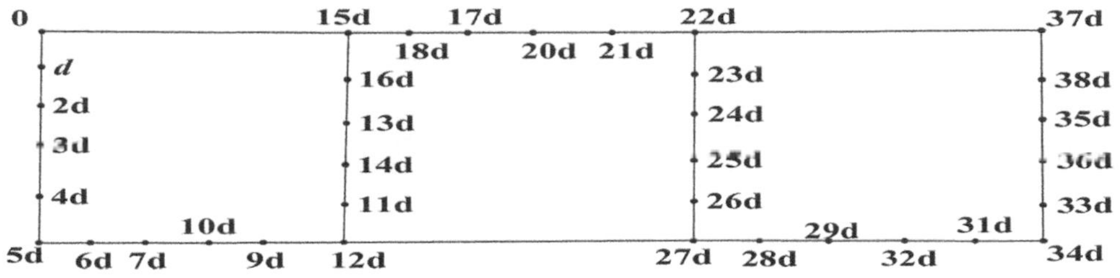


Figure 4. Graph $C_{16,6,3}$.

## 4 CONCLUSIONS

In this paper, on the basis of graph $C_{4k}$ and $C_{8,i,n}$ ($i=1,2,3,4$) to further expand, we give the labeling of graph $C_{4k,i,n}$, then we prove that the labeling is an arithmetic labeling in the way of mathematic induction, thus we obtain the graph $C_{4k,i,n}$ is $(d,2d)$—arithmetic graph and odd arithmetic graph. At last, we give the arithmetic labeling of the graph $C_{12,5,3}$ and $C_{16,6,3}$.

## ACKNOWLEDGEMENT

This study was supported by natural science foundation of China (11261019, 11126171) and natural science foundation of Jiangxi province (GJJ12309).

## REFERENCES

Achaya, B.D. and Hegde, S.M. 1990. Arithmetic graphs. Graph Theory, (14):275–299.

Ergen, L. and Dan, W. 2009. Two Graphs Arithmetic Labeling. Journal of East China Jiaotong University 26(5):89–92.

Ergen, L. and Dan, W. 2010. A Type of Arithmetic Labels about Circulationg Ring. CCTAIV IFIP AICT Vol. 347:391–398.

Kejie, M. 1991. Gracefulness. The press of technology, Beijing.

Liangzhi H. 2008. On Odd Arithmetic Graphs. Journal of Mathematical Research & Exposition.

*Information Systems and Computing Technology – Zhang & Gu (eds)*
*© 2013 Taylor & Francis Group, London, ISBN 978-1-138-00115-2*

# A dual-market Cloud model for Cloud computing service

Xiaohong Wu, Yonggen Gu & Jie Tao
*Huzhou Teachers College, School of Information and Engineering, Huzhou, Zhejiang, China*

ABSTRACT: Cloud computing presents innovative ideas for new internet services which utilizes the computing resource and storage resource to dynamically provide on-demand service for users. This paper focuses on dynamically relationship in supply and demand between users and service/resource providers, proposing a dual-market Cloud framework containing application layer, service layer, resource layer, in which the service market is separated from resource market in Cloud computing. The comprehensive description related to overall system architecture and its elements that form the Cloud dual-market model are introduced, and some research issues are discussed according to the analysis of the model.

## 1 INTRODUCTION

Cloud computing of computing as a utility, makes software even more attractive as a service and shaping the way IT hardware is designed and purchased. Developers with innovative ideas for new internet services no longer require the large capital outlays in hardware to deploy their service or the human expense to operate it. They need not be concerned about over-provisioning for a service whose popularity does not meet their predictions, thus wasting costly resources, or under-provisioning for one that becomes widely popular, thus missing potential customers and revenue (Vaquero L M, 2008).

Cloud computing is different from grid computing that focus on realizing resources sharing and integration. Cloud computing is more suitable for analysis of economics and market perspective than grid computing because of the core concept of on-demand service and commercial characteristics. Market-based approach to analyze Cloud computing service has been studied by some researchers (Buyya R, 2008; Tsai W T, 2010; Buyya R, 2010; Nallur V, 2010). Most of the research focus on the Cloud resources allocation, and economic-based approaches, scheduling and managing resources in Cloud computing are proposed on the basis of research on grid computing, so services and resources are put together. In paper (Hassan M M, 2009), an auction model for resources/service allocation is given, but cannot indicate the difference between resources allocation and service allocation.

Our proposed Cloud computing service model differs from the previous related approaches in two contexts. We introduce a three-layer federated Cloud architecture which includes Cloud application layer, Cloud service layer and Cloud resource layer. We also proposed Administrator Cloud dual-market model for Cloud computing service, in which the service market is separated from resource market in Cloud computing.

The rest of this paper is organized as follows: First, some related research is concisely introduced. Next, the comprehensive description related to overall system architecture and its elements that form the basis for constructing open federated Cloud is given. This is followed by some research issues in dual-market Cloud framework. Finally, the paper ends with brief conclusive remarks and discussion on perspective future research directions.

## 2 RELATED RESEARCH

InterCloud[5] is proposed by CLOUDS Laboratory of the University of Melbourne in order to build mechanisms for seamless federation of data centers of a Cloud provider or providers supporting dynamic scaling of applications across multiple domains in order to meet QoS targets of Cloud customers. InterCloud framework consists of Cloud exchange, client's broker and Cloud coordinator that support utility-driven federation of Cloud infrastructure: application scheduling, resources allocation and migration of workloads.

*The Cloud Exchange (CEx)*: CEx acts as a market maker for bringing together service producers and consumers. It aggregates the infrastructure demands from the application brokers and evaluates the available supply currently published by the Cloud Coordinators. It supports trading of Cloud services based on competitive economic models such as commodity markets and auctions.

*The Cloud Coordinator(CC)*: CC service is responsible for the management of domain specific enterprise Clouds and their membership to the overall federation driven by market-based trading and negotiation protocols. It provides a programming, management, and deployment environment for applications in a federation of Clouds.

*The Cloud Broker (CB)*: CB acting on behalf of users identifies suitable Cloud service providers through the Cloud Exchange and negotiates with Cloud Coordinators for an allocation of resources that meets QoS needs of users.

InterCloud framework is proposed by focusing on Cloud Infrastructure, and SaaS providers are consumers when the market means the resources exchange. But Cloud application service consumer will also have difficulty in selecting Cloud service providers in order to improve utility in meeting SLAs. Hence, it is necessary to build mechanisms for seamless federation of Cloud service providers for autonomic provisioning of services through different Cloud resource providers. Hassan (Hassan M M, 2009) proposed a Combinatorial Auction (CA) based Cloud resources/service allocation which can address the interoperability and scalability issues for Cloud computing, but it cannot indicate the difference between resources allocation and service allocation. The open Cloud model we propose separates the Cloud service from Cloud resources on the basis of InterCloud framework in order to analyze utility-Orient Cloud and their federation more clearly.

## 3 DUAL-MARKET CLOUD MODEL

The Cloud computing definition of NIST includes three different service models: SaaS (Cloud Software as a Service), PaaS (Cloud platform as a Service) and IaaS (Cloud infrastructure as a Service). The relations of three blocks—SaaS, PaaS and IaaS—should be analyzed from two aspects. One is from the point of view of user experience, relations between them are independent from this aspect, because they face different types of users. The other is from the technical point of view, they are not simple inheritance (SaaS based PaaS, and PaaS based IaaS), SaaS can be deployed on PaaS, or directly above IaaS, and can also be directly built on the physical resources above; PaaS can build on IaaS above, and can also be built directly on top of physical resources. So we will divide the market-oriented Cloud structure into three levels: the application layer, service layer and resources layer, and the proposed dula-market Cloud architecture as follows.

### 3.1 *Dual-market Cloud framework*

Dual-market Cloud framework consists of application layer, service layer and resource layer and two interfaces (resources market and service market).

Figure 1 shows the components of the market-based architectural framework consisting of Cloud service market, resource market and three layers.

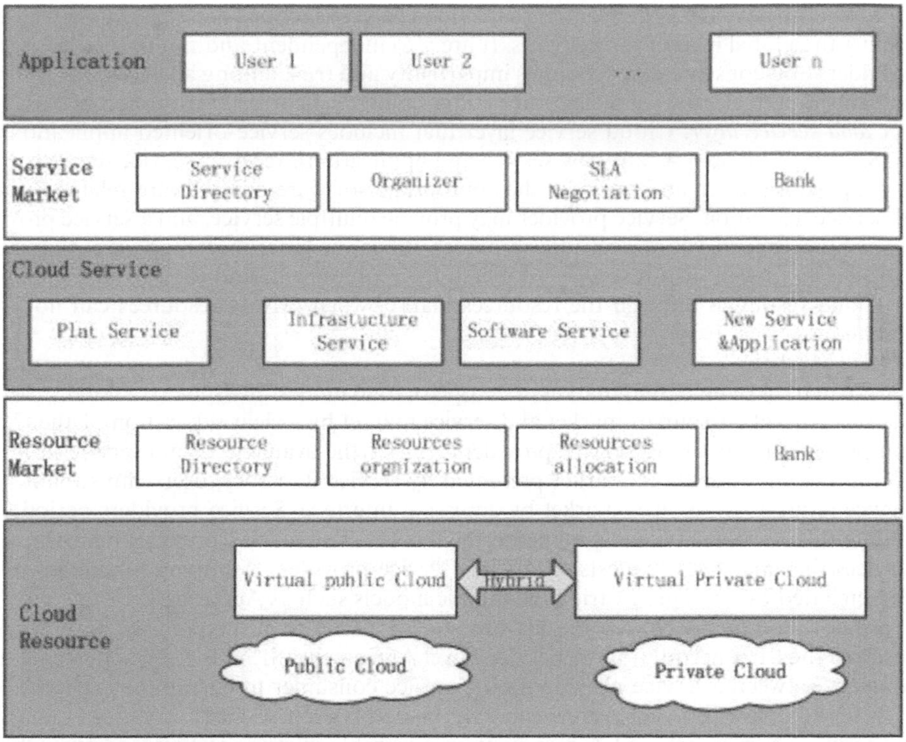


Figure 1. Dual-market Cloud framework.

1. Cloud Resource Layer
   Cloud resource such as computing servers, storage servers and network bandwidth in the bottom layer provided by resource owners can be divided into public Cloud resources and private Cloud resources from the perspective of whether the Cloud resources can be shared. Cloud resource can be virtualized by owners or third-party's virtualization technology. Virtual public Cloud and virtual private Cloud also can be associated into virtual hybrid Cloud. Although Cloud services and Cloud resources are in different layer, they are not necessarily separate completely. For example, a SaaS provider can support its own services provided to customers by his own private Cloud resources, so he is also in Cloud resource layer when he provides resource as resource owner.
2. Resource Market
   Resource market is the interface between Cloud resource layer and Cloud service layer, and it provides functions such as Cloud resources directory registering, resources dynamic bidding, resources combination, and resources payment management as discussed below.

   *Resource directory registering*: All available Cloud resources are registered in the market for lease. Cloud resource providers can publish the available supply of resources and their offered prices. The market directory allows the global participants to locate resources providers or consumers (service providers) with the appropriate bids/offers.

   *Resources organizing*: Resources market would organize the resources according to the demand. It can combine the available resources in the directory from single provider or different providers for requirement.

   *Resource allocation*: Resource market maker do not represent any providers or consumers as a third party controller who need to be trusted by participants in total control of the entire trading process. Resource allocator allocates resources by some economic model such as auction to improve utility and balance load.

*Bank*: The payment system enforces the financial transactions pertaining to agreements among the global market participants. It are also independent and not controlled by any providers and consumers; it promotes impartiality and trust among all Cloud market participants and makes the financial transactions conducted correctly without any bias.

*Cloud service layer*: Cloud service layer that includes service-oriented applications of IaaS, PaaS, SaaS and XaaS (new service and application), receives service requests from the application layer, and provides the appropriate software or hardware-related services by service allocation. Service provider may provide multiple service, and a service provider at the same time maybe is a resource provider. When the latter meets service requirement, it begins to find its own resources to match the requirement, and then requests for public Cloud resources through the resources market when private resources can not meet demand.

3. Service Market
   Service market is the second market in the open Cloud model, dual-market Cloud structure has two markets: resources market and service market by a clear separation of the Cloud computing service. Cloud service providers register the available Cloud service information in the Cloud services market presented by service directory, users can submit their service requests to services market by accessing interface. Service providers periodically update their availability, and pricing on the market. The market provides match-making services that map user requests to suitable service providers. Mapping functions will be implemented by leveraging various economic models such as Auction.

   Some functions in service market are similar to resource market, so here we only describe the SLA negotiation. A Service Level Agreement (SLA) is a formally negotiated contract between a service provider and a service consumer to ensure the expected level of a service. Service Level Agreements are made between services at various points of the service choreography. SLA may contain a variable number of Service Level Objects (SLOs), with variable target values. For example, SLO can be presented by{Cost, Response time, Resolution}.
4. Application Layer
   Users submit service requests from anywhere in the world to the the Cloud to be processed, and the user-oriented applications used to receive the user's service request, and deliver the service requests which come from different locations in the world to the service layer.

### 3.2 *Cloud dual-market model*

According to the Cloud framework, Figure 2 shows the Cloud dual-market model, the participants of model are described.

*Users*: Users is one participant of service market who submit service requests (IaaS, PaaS, SaaS) to the Cloud Service Market by Web service or software like a Web browser. Service request should presents the user's location, and consult Service Level Agreement (SLA) and compensation with service providers.

*Service providers*: Service Providers is another participant of service market, and at same time, it is one participant of resource market. On one hand, service providers publish their service information such as service types, pricing, etc; On the other hand, they submit their request to Cloud resource market by converting service request to the infrastructure requirements.

*Resource owners*: Resource Owners is one participant of resource market who publish the resources information such as number, location, price and bandwidth whose motivation is to achieve efficient resource utilization by maximizing revenue and minimizing the cost of resource consumption within a market environment.

Users and service providers meet in service market and negotiate for the sales of services by means of the signature of a Service Level Agreement. Resource owners get customers (service providers) in resource market by auction, and resource owners must guarantee some quality of service terms contained in the SLA by managing properly their resources. Dual-market

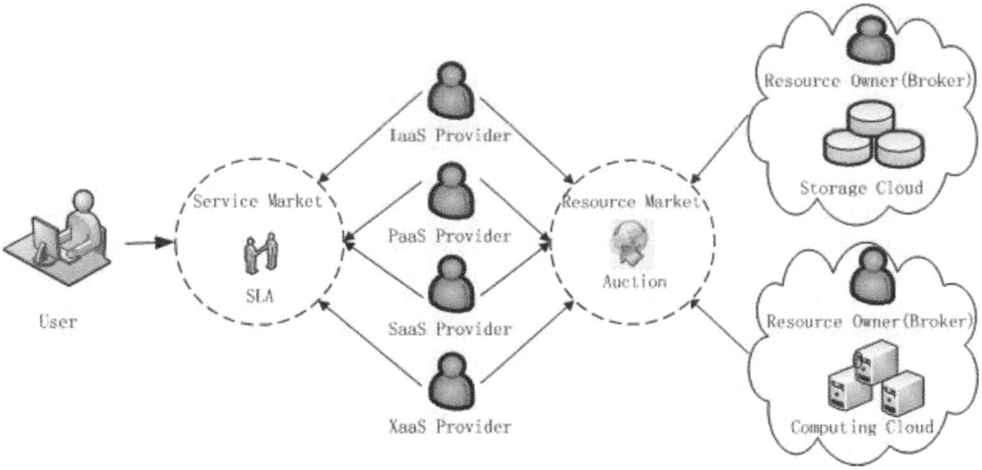


Figure 2. Dual market model of Cloud computing.

Cloud arises as an efficient way to allocate resources for the execution of tasks and services within a set of geographically dispersed providers from different organizations.

## 4 RESEARCH ISSUES

The diversity and flexibility of the functionalities envisioned by dual-market model, combined with the magnitudes and uncertainties of its components (workload, compute servers, services, workload), pose difficult problems in effective provisioning and delivery of application services. In particular, finding efficient solutions for following challenges is critical to exploiting the potential of federated Cloud:

1. SLA and the Target Transformation in Two Markets
   The SLA presents the service terms and conditions that are being supported by the Cloud service provider to respective user on a per user basis. For Cloud service users, Cloud computing is used to improve availability and system performance objective, which is the SLA must be addressed. From the application point of view, the performance of Cloud services is the sum of network performance, application performance and infrastructure performance. Clouds resources provider is responsible for the performance of infrastructure, Cloud service providers should be responsible for the three above.
   In this model, the resource provider is not committed to the high-level Service Level Objectives (SLOs) in SLA for the service (e.g., response time, throughput, etc.). Instead, the resource provider commits itself to an infrastructure SLA that governs the terms and conditions of capacity provisioning according to the explicit requirements of the service provider. The service provider supplies explicit capacity requirements and is charged for actual capacity usage in line with the "pay as you go" model. But how can present infrastructure SLA in resource market according to high-level Service Level Objectives which has consulted on service market between user and service provider?
2. Service Pricing Model of Service Provider
   In the service market, a service provider should be equipped with a price-setting mechanism which sets the current price for the service based on market conditions, user demand, current level of utilization of the service and the leasing cost of resource. Price can fluctuate dynamically with market environment but also can be relatively fixed during period of time. Service providers gain their utility through the difference between the price paid by the users for gaining service and the cost such as leasing resources for providing service

paid by themselves. For maximizing the effectiveness of the service providers in the case of SLA assurance, we would use a Multi-dimensional Optimization Problem in rational pricing of Cloud services. For solving the MOP, we should explore multiple heterogeneous optimization algorithms, such as dynamic programming, hill climbing, parallel swarm optimization, and multi-objective genetic algorithm.

3. Mapping of Services to Resources
   The process of mapping services to resources is a complex undertaking, as it requires the system to compute the best software and hardware configuration (system size and mix of resources) to ensure that QoS targets of services are achieved, while maximizing system efficiency and utilization. This process is further complicated by the uncertain behavior of resources and services. Consequently, there is an immediate need to devise performance modeling and market-based service mapping techniques that ensure efficient system utilization without having an unacceptable impact on QoS targets.
4. Economic Models Driven Optimization Techniques
   The market-driven decision making problem is a combinatorial optimization problem that searches the optimal combinations of services and their deployment plans. Unlike many existing multi-objective optimization solutions, the optimization models that ultimately aim to optimize both resource/service-centric (utilization, availability, reliability, incentive), user-centric (response time, budget spent, fairness) QoS targets need to be developed.

## 5 CONCLUSIONS

This paper presents a new dual-market Cloud framework and model, in which service market and resource market are proposed by separating Cloud service from Cloud resource. This framework facilitates to analyze Cloud computing market in large, federated and highly dynamic environments. This will provide enhanced degrees of scalability, flexibility, and simplicity for management and delivery of Cloud services and Cloud infrastructure in federation of Clouds.

## ACKNOWLEDGEMENTS

The research is supported by the National Nature Science Foundation of China (No. 61170029) and the Zhejiang Natural Science Foundation of China (NO. Y1111000).

## REFERENCES

Buyya R, Yeo C S, Venugopal S. Market-oriented cloud computing: Vision, hype, and reality for delivering it services as computing utilities[C]//High Performance Computing and Communications, 2008. HPCC'08. 10th IEEE International Conference on. Ieee, 2008: 5–13.

Buyya R, Ranjan R, Calheiros R N. Intercloud: Utility-oriented federation of cloud computing environments for scaling of application services[M]//Algorithms and architectures for parallel processing. Springer Berlin Heidelberg, 2010: 13–31.

Hassan M M, Song B, Yoon C, et al. A novel market oriented dynamic collaborative cloud service infrastructure[C]//Services-II, 2009. SERVICES-2'09. World Conference on. IEEE, 2009: 9–16.

Nallur V, Bahsoon R. Design of a market-based mechanism for quality attribute tradeoff of services in the cloud[C]//Proceedings of the 2010 ACM Symposium on Applied Computing. ACM, 2010: 367–371.

Tsai W T, Sun X, Balasooriya J. Service-oriented cloud computing architecture[C]//Information Technology: New Generations (ITNG), 2010 Seventh International Conference on. IEEE, 2010: 684–689.

Vaquero L M, Rodero-Merino L, Caceres J, et al. A break in the clouds: towards a cloud definition[J]. ACM SIGCOMM Computer Communication Review, 2008, 39(1): 50–55.

*Information Systems and Computing Technology – Zhang & Gu (eds)*

 *ISBN 978-1-138-00115-2*

# Technical requirements for integrated electronic information system in Big Data Era

Shi-liao Zhang, Ya-bin Kan & Peng Liu
*Operation Software and Simulation Research Institute, Dalian Naval Academy, Dalian, China*

ABSTRACT: Big Data Era (BDE) is coming, which plays an important role in technical requirements of producing, saving, transmitting and using data. In order to study the development of IEIS by the technological change of BDE, this paper studies the characters and technical value of BDE first, then analyzes the technical requirements of the Big Data for every function of IEIS. The research conclusions are important to develop and improve IEIS with times and creativeness of technical, which is significant meaningful for military.

## 1 INTRODUCTION

With the fast development of movable internet, the increasing use of cloud computing, the net of things and new application services, the flowing data has mushroomed in the internet, which make us be in BDE. At the same time, bigger military platform, integrated operation and system confrontation demand all persons and military equipment in one net named military net of things, which produces huge data, so Integrated Electronic Information System (IEIS) is also in BDE. If ISIE wants to work steadily, reliably and efficiently, ISIS must develop the techniques for big data producing, saving, transmitting and using.

## 2 BIG DATA CHARACTERS AND TECHNICAL VALUES

### 2.1 *Big Data characters*

Big Data has four characters such as Volume Big, Variable Type, Velocity Fast and Value High & Low Density, "Four V" for short.

### 2.2 *Big Data technical values*

The vast potential for future using of Big Data will evoke technology booming of its producing, saving, transmitting and using.

1. Improving the ability of producing data
   All the world has paid attention to application prospect of Big Data and demands precise and safe data timely from data terminal, which makes the information searching, catching, taking and screening be important.
2. Improving the ability of saving data
   "Four V" characters describe that the speed of producing data is very fast and the quantity is big, which demand the saving speed must be fast and the room must be adequate.
3. Improving the ability of data mining technique
   If you want to find useful information from the "Four V" Big Data, you must develop dynamic data mining technique.

4. Improving the abilities of analyzing data statistics, relation and forecasting acts
   In BDE, not only the technique of using data is developed, but also the technique of statistical analysis will be used to taking the characters of targets. The writer of "bursts", the most famous leader in Complexity Science of the world, named Albert-laszlo Barabasi, believes that 93 percent of acts of human can be forecasted after reviewing and verifying the specific conception of history. So, we believe that it is possible that we can forecast acts timely and precisely based on study on data statistics, relation and judging.
5. Improving the abilities of net fraction of coverage and data transmitting
   In BDE, moveable internet and the net of things are the medium of transmitting. So the net fraction of coverage is related to collect data, and the net speed is related to transmitting data. In order to make good use of Big Data, it is important to develop the net coverage rate and the technique of transmitting.

## 3 ANALYZING THE FUNCTIONS OF IEIS IN RELATION TO BIG DATA

### 3.1 *The main functions of IEIS*

IEIS is composed of all kinds of electronic information systems, which is an organic whole and accord with a unified architecture. IEIS mainly includes the following functions:

1. Command and control functions
   Commander brings the combat plan to practice by commanding, coordinating and controlling all persons, equipment and processes. The input of commanding and controlling is battlefield situation information. The output is decision results in time, which will be used to commanding and controlling.
2. Intelligence reconnaissance function
   The function is mainly to collect enemy's all intelligence by a variety of means, and the intelligence will be processed, analyzed, saved and distributed.
3. Early warning detection function
   The function is mainly to find targets information by a variety of means, and the information will be processed, saved and distributed.
4. Military communications functions
   This function is mainly to provide method and means for transmitting information, including line-of-sight communications, wireless communications, WAN communications. Communication contents include voice, data and video.
5. Electronic warfare and information warfare functions
   Electronic warfare is composed of all military acts in relation to the use of electromagnetic energy, and controlling electromagnetic spectrum to attack enemy, including electronic attack, electronic protection, electronic support, etc. Information warfare, including information attack and protection, is composed of all military acts to obtain information superiority by influencing the enemy's information processing based on information process, information system and network.
6. Military navigation and positioning function
   This assures platform to get its position, and guide a platform to a given location.
7. Integrated information support function
   This processes and sends all useful information for combat including surveying and mapping, weather, the process of operation, engineering, chemical defense, logistics, equipment, and electromagnetic frequency management.
8. Information security function
   Mainly includes the communications security, computer security and computer communication network security, etc.
9. Military information sharing and resource sharing function
   This collects all information from all military sensors, then all information is integrated treatment, saved and distributed by military information infrastructure.

### 3.2 *Functions in relation to producing Big Data*

In IEIS, there are kinds of information, including information accessed directly and indirection information produced by statistics, analyzing and prediction.

1. Information, including enemy commander personality traits and operation information, collected for a long time by intelligence personnel, satellite, etc.
2. Information of my commander personality traits collected and completed for a long time.
3. Documents and materials of our operational principles.
4. The first-hand nature data obtained by the meteorological, hydrological, time and other test equipment.
5. The first-hand information obtained by personnel who collects intelligence or reconnoiters intelligence, and surveillance equipment and sensors.
6. The first-hand information, including weapon system status and using record, obtained by data collection system of military.
7. Information of result by analyzing, calculating, and predicting the collected hydro meteorological data.
8. Information of navigational data and moving parameter calculated.
9. Information of tasks and plans according to the situation of the enemy, my feeling, environment analysis, etc.
10. Information of target attributes, position, moving parameter produced by the calculating of data fusion, situation assessment, treat level judged, etc.
11. Information of commanding orders, decision answers, control orders and operation effect, and so on, produced by commander.
12. Information of operational effect evaluation.

According to the above analysis, functions of IEIS related to producing Big Data are command and control, intelligence and surveillance, early warning detection, military navigation and positioning, electronic warfare and information warfare and military information sharing and resource sharing.

### 3.3 *Functions in relation to saving Big Data*

ISIE has three using patterns, including training, operation and replay, which must use a large number of data. So, all the data must be saving, and all functions is related to saving Big Data.

### 3.4 *Functions in relation to transmit Big Data*

ISIE must transmit all types of data to several domains, including land force, forces on the water and underwater, air force and space force. All information data should be transmitted by military communication network.

### 3.5 *Functions in relation to using Big Data*

For ISIE, the process of working for itself is the process of using all type of data, so all functions excluding functions in relation to producing and transmitting Big Data are belong to functions in relation to using Big Data.

## 4 TECHNICAL REQUIREMENT OF ISIE IN BDE

### 4.1 *Technical requirement for producing Big Data*

#### 4.1.1 *Technical requirement for command and control*

The data include the results of battlefield situation evaluation, calculating target threat level, distributing weapons to targets, orders of control and channel organizing of using weapons.

Face to the battlefield, called Big Battlefield, including a large number of targets, the following techniques should be developed quickly.

1. Quick and accurate evaluating technology of Big Battlefield situation
   Face to the large number of targets, IEIS should give the intent of every target, such as feigned activity, screen and real using weapons. This technique should recognize the intent of every target in Big Battlefield quickly and accurately.
2. Quick and accurate calculating technology of target treat level in Big Battlefield
   Face to the large number of targets, commander should know the treat level of target to me, which will help commander do the later work rightly.
3. Quick and accurate technology distributing weapons to targets
   In Big Battlefield, commander distribute limited number of weapons to such number of targets, which demand IEIS give a reasonable scheme allowing a weapon to a fixed target.
4. Quick and accurate technology of remote control
   This technology is to decide when to give target parameter, when and how to fire. In BDE and under the situation of military net of things, weapons are controlled remotely through network. So this technology is very useful.
5. Quick, flexible and efficient technology of organizing channel for using weapons
   In BDE and under the situation of military net of things, all equipment including sensors and calculators belong to one net. So, in such a big net, it is a complex system engineering. In order to using weapons efficiently, this technology is very important.

4.1.2 *Technology of producing data of intelligence reconnaissance*
Intelligence reconnaissance is a work keeping in the background for a long time, which relies on cooperation of many workers. Intelligence includes forms and deployment of forces, characters of commander, characters of operation, etc. In order to produce these variant data distributing widely, the following technology must be high efficient.

1. Technology of tracing and reconnaissance in space
   Every military activity was the result of training at ordinary times, and every using tactics is the specific implementation of the tactical principle, and every decision is the inertial application of training results. So, collecting these intelligence is very important to our commander making decisions. In order to spy on every military activity in widely space, satellite is the only kind of equipment who can do this. So, developing the technology of tracing and reconnaissance in space must be done rapidly.
2. Establish intelligence reconnaissance systems in the land, on the water and underwater, in sky and in space
   In information era, it is an important tactics principle to find enemy before enemy. With the developing of military net of things, a large number of sensors can cooperate well. So, we should establish intelligence reconnaissance in any reachable space, which will produce Big Data all time remotely, and give us good chance to find enemy before enemy.

4.1.3 *Technology of producing data of early warning detection*
The goal of early warning detection system is to as early as possible to find the enemy, which will give commander enough time to make decisions. Face to the high-tec partial war, the information is a very important element according to the principle of finding enemy as early as possible.

1. Developing technology of observing by satellite in space
   Because of the distance from the earth to satellite, satellite can spy on a very large area, and many satellites can spy on the whole earth. So, many satellites spying on and several satellites calculating can give large number of useful data. The technology how to organize all these satellites to work well and how to give these efficient data demand to strive to develop.

2. Developing technology of observing by self-platform
   Observing of self-platform is the deadline of find information of enemy, so this is the basic protection of data. This technology must be developed quickly.

4.1.4 *Technology producing data of military navigation and positioning*
Operational requirement for navigation and positioning is fast, accurate and stable, which ensure commander and weapon system can have sustainable access to timely and accurate location information. It is the inevitable requirement of military navigation and positioning that comprehensive use of satellite navigation and positioning, prediction of self-platform, astronomical calculation and physiographic calculation.

4.1.5 *Technology of producing data of electronic warfare and information warfare*
Electronic warfare mainly completes the task of specific electronic counter-measures, and information warfare mainly completes the task of specific counter-information. Face to the battlefield with complex electromagnetic environment and large number of targets, the follows technologies are important.

1. Technology producing data of electronic information by reconnaissance
   In complex battlefield, IEIS can observe and collect electronic information of enemy. These information is useful to disturb enemy's equipment rightly.
2. Technology producing data of information warfare is similar to intelligence reconnaissance

4.1.6 *Technology of producing data of integrated information support*
Integrated information support is difficult in any domain of IEIS. If IEIS wants to have enough information, it has to establish large number of sensors in a very broad space. The follow technology is necessary.

1. Constructive technology of sensors system;
2. Technology of tracing and analyzing information;
3. Technology of integrating and predicting information of sensors;
4. Technology of collection and automation of sensors system.

4.2 *Technical requirement for saving Big Data*

The Big Data is not only used to operation but also used to replay, analyze, study and collect long time for regulation research later. So, saving Big Data is an important work, and several technologies should be developed.

1. Quick compressing and predicting technology of supersize data;
2. Quick saving and predicting technology of supersize data.

4.3 *Technical requirement for transmitting Big Data*

In BDE, information transmitted in military net of things has many types, such as not only voice, data and fax but also battlefield images, moving images of forces and electronic warfare data, so volume of nodes must be enough and transmission speed and precision rate must be high. According to the demanding of integrated operation, mobile operation and high speed self-organization, we should develop the following technology excluding increasing the bandwidth.

1. Military communication technology of space-based for extra large number of nodes.
2. Area communication technology with reorganization and high speed.
3. Mobile military communication for extra large number of nodes and joint operation.

4.4 *Technical requirement of using Big Data*

4.4.1 *Technical requirement for command and control*

The input of command and control system are the information of large number of targets, the information of battlefield situation, the results of calculating target treat level, the results of distributing weapon to targets, orders of control and channel organizing of using weapons.

4.4.1.1 Technology of using the information of battlefield targets

Face to the large number of targets, IEIS should still work well, which demands IEIS doing the followings.

1. Screening technology to pick up the real and useful data from the information of the large number of battlefield targets and discard unreliable data.
2. The technology of cluster and data mining to pick up the data for each target.
3. The dynamic calculating technology of quick data fusion and getting the precise information timely.
4. The technology of analyzing the dynamic history data for correlation and prediction.

4.4.1.2 Technology of using data of battlefield situation

Battlefield situation is one of the proofs of next step such as doing operation plan and using weapons. So, IEIS should use the information of battlefield situation to make a prediction and recognize the intent of every target.

4.4.1.3 Technology of using data of target treat level

Target treat level is one of the proofs of using weapons, which will be used to decide when and how to using weapons.

4.4.1.4 Technology of using the results of distributing weapons to targets, orders of command and control and channel organization of using weapons

These tree types of information is directly used. In operation, commander use sensors, command and controls system and weapons according to the results of calculated channel, and decide how many and when weapons are used in accordance with the results of distributing weapons and orders of command and control.

4.4.2 *Technology of using data of electronic warfare and information warfare*

The work efficiency of electronic warfare has more to do with the collection and study on the enemy' electronic information. The data of information warfare is scattered, but all information is strategically useful.

In practice, if commander wants to make a good decision on the electronic warfare and information warfare with right attack or defense, current information and history must be just right for a match.

There are several technologies that should be studied quickly and well.

1. Collecting electronic information of the smallest operation unit of enemy.
2. Establishing the enemy's parameter library of tactical and technical characteristics by data mining, cluster, rational analysis and prediction.
3. Based on the history information of commander's characteristics and platform, predicting the tactical acts of enemy by data mining, cluster and rational analysis.
4. Based on all kinds of information from different ways, predicting the intent of enemy by rational analysis, and help make decision on information attack and defense.

4.4.3 *Technology of integrated information support*

Information support has two demands. One is that the library has accumulated enough information, and the other is that information can be send to army timely.

1. Scientific and reasonable management technology, which ensures the data from a large number of nodes are clear to use.
2. The technology to send specific useful data to a given army rightly.

3. Based on the long time using data, analyzing the using characteristics.
4. Optimal support technology, which will make the limited ability of support match the large number of used nodes.

4.4.4 *Technology of sharing the information and resources*

Face to the extra large number of information data and different users with more defined professional field, the precise of share information to certain target is very important. So the technologies, including data mining, cluster and prediction, should be developed based on the history information, which will ensure sending right data to right target.

## 5 CONCLUSIONS

From the above, we know that every stage of the whole cycle of Big Data, including producing, saving, transmission and using, demands for developing several technologies. At producing stage, this article points out that collecting and accumulating data long time are needed. At saving stage, this article tells us that the dynamic fast-compression and fast-saving are very important. At transmission stage, this article draw the conclusion that rational transmitting technology must be developed. At the using stage, this article gives us the idea that developing these technologies, including dynamic Big Data mining, cluster, analyzing, reasoning and predicting, is requisite.

## REFERENCES

Altert-Laszlo Barabasi. 2012. BURSTS: The Hidden Pattern behind Everything We Do.
Chen Ruming. 2012. Challenge, value and coping strategy in Big Data Era(17):14–15.
Cheng Shian. 2012. Net management in Big Data Era focused on search engine. observer:15.
Chief Editor. 2012. 11. 11. Meeting the coming of Big Data Era. China Telecommunications Trade:6.
Searching for mechanism of war within Big Data. www.ciotimes.com/industry/js/78–575.html.
Yuan Jie. 2012. Commercial opportunity in Big Data Era. Modern Commerce:10–13.

*Information Systems and Computing Technology – Zhang & Gu (eds)*
*© 2013 Taylor & Francis Group, London, ISBN 978-1-138-00115-2*

# The research of CRM based on web mining approach

Junhua Ku
*Department of Information Engineering, Hainan Institute of Science and Technology, Haikou, China*
*School of Computer Science, China University of Geosciences, Wuhan, China*

Na Chang & Wei She
*Department of Information Engineering, Hainan Institute of Science and Technology, Haikou, China*

Zhihua Cai
*School of Computer Science, China University of Geosciences, Wuhan, China*

ABSTRACT: Web mining has undergone tremendous transformation from an emerging technology and a new scientific curiosity to a well-established discipline having a well-defined and established body of knowledge in the near future. It is a good time to think about novel promising areas that will advance Web mining in the near future. In this paper, we believe that CRM is such an area that can benefit from and contribute to further advancement of Web mining research. In this paper, we discussed how and why Web mining is an important component of CRM and how some of its techniques can help advancing the field of CRM further. At the same time, CRM problems can also impose certain additional requirements on Web mining which may lead to the development of novel Web mining methods.

## 1 INTRODUCTION

Web mining has undergone tremendous transformation from an emerging technology and a new scientific curiosity to a well-established discipline having a well-defined and established body of knowledge. While celebrating all this progress, it is a good time to reflect on the next growth area in Web mining and identify either a big application domain or a novel set of methods that can significantly advance the field in the near future (Adomavicius G & Tuzhilin A, 2001). We believe that it maybe the CRM which is such an area that can benefit from and contribute to further advancements of the Web mining research.

As discussed in many articles, data and Web mining methods can be used extensively across the whole spectrum of CRM problems, ranging from the issues of getting, keeping and growing the customers, better estimation of performance metrics, such as Lifetime Value (LTV), estimating customer responses, collecting and organizing customer data, and building customer profiles and user models.

Since CRM systems differ from the classical Web-based applications, it is important to examine if the traditional Web-based methods used in those applications can be directly applicable to the CRM problems. If they are not, then the next question is whether the existing Web mining methods need to be "tweaked" or fundamentally reconsidered and novel Web mining techniques developed that are more suitable for the CRM problems. For example, consider the problem of estimating unknown ratings in recommender systems. One of the techniques for doing this is content-based filtering that recommends those items that the user liked in the past. Its solution typically involves developing a classifier that predicts an unknown rating based on the characteristics of the items that the user liked before.

One example of these past "characteristics" in some online applications is browsing and navigational activities performed by the user on a website, and, clearly, Web mining plays an important role here. Since recommenders and CRM systems constitute a novel and underexplored class of Web mining applications, it is not clear if the current Web mining methods can be directly applied to these problems. This issue needs to be explored further by the Web mining, CRM and recommender systems communities with the goal to see if the existing Web mining approaches are directly applicable to these problems or novel methods need to be developed in those cases when they are not directly applicable. This could be a novel research direction for the Web mining community. Finally, such studies should not be applicable only to the recommender systems problems but can be directed towards other areas of CRM.

In this paper, we discussed how and why Web mining is an important component of CRM and how some of its techniques can help advancing the field of CRM further. At the same time, CRM problems can also impose certain additional requirements on Web mining which may lead to the development of novel Web mining methods. This article is organized as follows. The next section briefly reviews basic understanding of CRM. Sect. 3 describes the connotation of web mining approach. In Sect. 4 we will present future research directions for CRM from our perspective. Finally, we will draw the final conclusions in Sects. 5.

## 2 WHAT IS CRM

### 2.1 *Key concepts and definitions*

Customer Relationship Management (CRM) comprises a set of processes and enabling systems supporting a business strategy to build long term, profitable relationships with specific customers (Ling C & Li C, 2001). Customer data and Information Technology (IT) tools form the foundation upon which any successful CRM strategy is built. In addition, the rapid growth of the Internet and its associated technologies has greatly increased the opportunities for marketing and has transformed the way relationships between companies and their customers are managed (Ngai E et al., 2009). Since CRM still constitutes a growing field, there still exist different points of view on what it is, as expressed by the academics and the practitioners. Some representative definitions of CRM proposed in the marketing literature are:

1. The goal [of CRM] is not just to offer excellent products and services but to get, keep and grow the best customers (Kotler P, 2004).
2. CRM is an enterprise approach to understanding and influencing customer behavior through meaningful communications in order to improve customer acquisition, customer retention, customer loyalty, and customer profitability. CRM is an iterative process that turns customer information into positive customer relationships (Swift R, 2001).
3. The aim of CRM is to identify, acquire, serve, extract value from, and retain profitable customers by interacting with them effectively in an integrated way across the full range of customer-contact points for marketing, sales and service, including via e-mail, mail, phone, over the Web, in person, and so on (Geoffrion A & Krishnan R, 2001).
4. CRM involves treating different customers differently (Peppers D & Rogers M, 2004).

Although different, all these definitions identify several important points about CRM. Collectively, they maintain that the purpose of CRM is to *get* (identify and acquire), *keep* (serve and retain) and *grow best* customers.

### 2.2 *The classifications of CRM*

Form the view of the structure of the discipline, To get, keep and grow the best customers requires extensive functionality on the part of a CRM system that is usually divided into *operational* and *analytical* CRM (Peppers D & Rogers M, 2004). Operational CRM is broadly characterized by various activities that the firm has to *do* to make customers more valuable. These activities usually include front office business processes supporting the entire range of customer interactions from identifying potential customers to sales and customer support. They critically depend on the customer contact history that records all the interactions with the customer across various contact channels, including the Web, phone, email, FAX, mail and in person and provides integrated "view" of the customer.

Analytical CRM is characterized by various things that a firm has to *know* about the customers to make them more valuable. This includes the analysis of customer data created by operational CRM and other sources that is done in order to understand (1) the customer needs, interests and desires, (2) how happy the customer is with the services provided by the firm, including estimating probabilities of customers' defections, (3) how valuable the customer is for the company, and (4) how to "grow" the customers to make them more valuable.

## 3 THE CONNOTATION OF WEB MINING APPROACH

World Wide Web is serving us information, in the order of hundreds of terabytes and is expanding rapidly. As the web is largely unrecognized with vast information, collecting and analyzing the data is very difficult. Web mining techniques are used to extract knowledge out of hundreds of web sites. Knowledge acquisition is used for activities in CRM. Accurate Web usage information helps customer management, improve cross marketing/sales, effectiveness of promotional campaigns, and find the most effective logical structure for the web space and so on. Web Mining is divided into three distinct categories (Etzioni O, 1996) e.g., Web Content Mining, Web Structure Mining, and Web Usage Mining, according to the kinds of data to be mined.

*Web Content Mining* is the process of extracting useful information from the contents of Web documents. Content data corresponds to the collection of facts in web page. It consists of text, images, audio, video, or structured records such as lists and tables. Text mining and its applications have been the most widely researched topics.

*Web Structure Mining,* the structure of a typical web graph consists of web pages as nodes, and hyperlinks as edges connecting between two related pages. It is the process of discovering structure information from the Web. This is further divided into two kinds, based on the structural data used such as *Hyperlinks*, *Intra-Document Hyperlink* and *Document Structure.*

*Web Usage Mining* discovers interesting usage patterns from web data for analysis of customer profiles on web-based applications. Usage data captures the identity or origin of web users along with their browsing behavior in a web site. It is further classified depending on the kind of usage data i.e., *Web Server Data, Application Server Data* & *Application Level Data.*

## 4 WEB MINING APPROACH IN CRM

CRM is a very broad field having many interesting and underexplored research problems, ranging from the development of novel methods of getting, keeping and growing customers to the identification and studying novel performance measures to the development of better customer profiling and modeling methods. In this chapter we will focus only on some of the research directions that can be of highest interest to computer scientists and that are the most important to the field of CRM from our perspective. These problems mainly include the issues of customer segmentation analysis, computational methods for getting, keeping and growing the best customers, customer feedback mechanisms, customer profiling and modeling, data organization, administration and governance. We will start these discussions with the customer segmentation analysis in Sect. 4.1. We will also discuss potential applications of data mining in CRM throughout Sect. 4.

### 4.1 *Customer segmentation analysis*

Customer segmentation is the foundation of firms' effective sales, marketing, and services. A lot of customer data can be efficiently selected through the web mining analysis to reflect the characteristics of customers attributes, such as gender, age, occupation, position, education degree, annual salary, the average consumption and so on. According to these characteristics signs, firms can divide customers into several types of consumption value, such as high value customer, medium value and elementary value customers and so on. Also, through the analysis of web mining the firms can obtain all kinds of information about customers, such as interests, habits, tendencies, requires and tendency, etc. Finally, As a part of building "learning relationships" and customer "development," the firm interacts with the customers in many different ways. These interactions consist of specific types of *actions* taken by the firm at certain times, and some examples of these actions, such as recommending products and services, presenting or sending information to the users, sending a product to a customer, subscribing a customer for some service, connecting a customer with another person, etc.

Firm can segment the customer base into several segments and build user models for each segment. This approach has been advocated and studied extensively in marketing (Wedel M & Kamakura W, 2000). In their research, however, marketers rely on certain paradigms developed in that discipline that are usually not scalable to large modern databases. Therefore, it is important for Web data mining researchers to re-examine these and develop new alternative computational approaches to customer segmentation that better suit specific CRM problems. One important issue is the topic of micro-segmentation: how to split a large customer base, such as all of the China Telecom customers, into a large number of micro-segments, how to build user models for very many segments and how to manage a large collection of these models effectively. Some initial approaches to this problem have been developed in (Jiang T & Tuzhilin A, 2006), but much more work is needed to produce a deeper understanding of this problem. Another way to segment customer bases is via identification of communities in social networks via the community discovery and mining methods that allow identifications of similar users (i.e., belonging to the same community) and building user models of the members of the identified communities.

### 4.2 *Computational methods for getting, keeping and growing the best customers*

The issue of getting, keeping and growing the best customers addresses the main question of CRM and, therefore, lies at the heart of the field. The topic of customer acquisition in CRM applications has received only a cursory attention in the CRM literature, despite its importance. In particular, the customer acquisition problem in CRM is applicable only to the customers and has two flavors: when customers are anonymous (and the firm does not have any personal and demographic information about them) vis-à-vis the case when the customers have registered with the firm and have already provided at least some of the profiling information in the form of the demographic and psychographic data about them. Finally, the goal of customer acquisition in CRM is to get those new customers for whom the firm can build "learning relationships". In short, customer acquisition is important for CRM researchers to make further study. One way to approach this problem, at least in the case when customers have already registered with the firm and the firm has some of the demographic and other profiling information about the customers, is to apply clustering methods, find the most appropriate segments for the new customers and deploy the same actions that the firm is applying to the similar and already established old customers. However, this approach is less suitable for the anonymous customers who have not yet registered with the firm.

The important computational research stream related to CRM is the area of Web analytics. Web analytics focuses on making website visitors to get engaged in certain activities on the site or take certain actions, such as clicking on certain objects and buying certain products, and subsequently measuring results of these activities. There exists a very extensive body of research in Computer Science studying various aspects of this problem, including numerous publications in most of the past Web Mining and Web Usage Analysis workshops. This work differs, however, from the CRM-related research because CRM work focuses primarily on the *iterative* process of *repeated* interactions with the customers with the goal of improving customer long-term experiences over time and maximizing long-term performance measures, such as the LTV and CE. Nevertheless, it is important to find the balance between the longer-term and the shorter-term objectives of the CRM and the Web analytics approaches and determine the ways in which each of the research streams can benefit from each other. Since only limited information exists on such anonymous customers, various browsing and other online activities become crucial for building good predictive models for these customers. Therefore, the CRM problem of getting new customers relies, to a large extent, on Web mining techniques, and deploying good Web mining methods becomes crucial for developing robust solutions to this problem.

### 4.3 *The customer feedback problem*

After the firm determines the most appropriate set of actions for a customer and launches a campaign based on this determination, the customer reacts to the campaign with various types of *responses* according to the action-response model. Some of the types of responses to the actions of the campaign include customer (1) Ratings of individual or groups of actions in a campaign, including of recommended products and services, (2) Purchases made as a result of various actions of the campaign, (3) Click-through's in response to certain actions (such as display ads or recommendations), and (4) Reviews of products and services published by the customer and various types of opinions contained in these reviews that can be mined using opinion mining and sentiment analysis methods (Pang B & Lee L, 2008).

Collectively, all these customer responses constitute *customer feedback* to the campaign actions launched by the firm. These feedback responses affect the values of the customer— and firm-centric performance measures, the values of which need to be recomputed based on this customer feedback. This re-computation completes the action-response cycle, and the firm is ready to launch the next campaign for customer based on the new values of the performance measures.

Some of the main research questions pertaining to the customer feedback problem are:

1. How to incorporate various types of customer responses listed above, such as rating specifications, click-throughs and purchases, into one overall performance measure, such as LTV and CSM.
2. What are the exact mechanisms of adjusting the customer-centric and the firm-centric performance measures, such as LT V and CSM, based on the customer feedback.
3. Customer responses constitute important data that can be used for labeling cases that are subsequently used for supervised learning purposes. Therefore, the problem of active learning becomes important, i.e., which customer feedback information is the most critical in a CRM application and what is the best way to obtain it. This connection to active learning in DM should be explored further.

There is a diverse set of research issues and problems. However, all of them depend critically on one fundamental problem of "deep" understanding of customers, which lies at the heart of any CRM system. We will discuss this problem in Sect. 4.4.

### 4.4 *Customer profiling and modeling*

Identification of the most appropriate actions on the part of the firm is crucial for getting, keeping and growing the best customers, as we argued throughout the paper. Furthermore, identification of these most appropriate actions depends, to a large extent, on a good "understanding" of the customers that arises from building comprehensive profiles of customers and accurately modeling their behavior.

The topics of customer profiling and user modeling has been extensively studied in data and Web mining (Ngai E et al., 2009; Nasraoui O et al., 2008) However, it is important to extend this work and apply it to the CRM problems. One such attempt was done by Nasraoui et al. (Nasraoui O et al., 2008), where the authors describe how to mine evolving user profiles and use them in recommender systems in order to build relationships with the customers. Moreover, the approach presented in Nasraoui et al. (Nasraoui O et al., 2008) was tested on a real-life case study of the National Surface Treatment Center portal that maintains extensive information about corrosion and surface treatment. More work is needed in this direction to advance our understanding of how to build and use customer profiles and develop user models for solving CRM problems.

Some of the interesting research directions along these lines are:

1. *Connecting customer profiles and user models with action ability.* Despite its importance, action ability still remains an underexplored concept in data mining, personalization and CRM. However, it is crucial to build customer profiles and user models so that the CRM systems would be able to take the most appropriate actions, and the CRM researchers should work on studying this connection.
2. *Correct vs. incorrect understanding of customers.* Misunderstanding and misinterpretation of prior customer actions or drawing premature conclusions about customer activities can lead to disastrous results, ruining the relationship and hard earned trust of the customer (Padmanabhan B et al., 2001). In other words, one disastrous action on the part of the firm can significantly increase the risks of customer defection. Therefore, it is crucial to avoid these types of misunderstandings by developing robust and correct models of customers. This topic has been under-explored in the CRM community and needs to be studied much more extensively.
3. *Questioning prior knowledge.* Prior knowledge about customers can come from different sources and some of this prior information can be conflicting for various reasons. For example, customer preferences can evolve over time and may not match the old preferences still stored in the customer profiles. For example, at some point a customer may express interest in books on entrepreneurship while running a startup company. A few years later, the customer could have moved into a different business and lost the interest in the startup books. This change in preferences should be reflected in the customer's profile. Another reason is that customer preferences can be different in different contexts, and it is necessary to reconcile these differences. For example, a customer may prefer watching action movies on weekdays and dramas on weekends. If this contextual information is not reflected in the user model of that customer, then one profile may contain action movie preferences and another profile (coming from a different source)—drama movies. These two profiles, obviously, need to be reconciled to create a more complete picture of that customer's preferences. Therefore, an important research problem is to define and identify these types of conflicts and develop mechanisms for resolving them. Another research problem is the determination of robust updating policies of customer profiles in order to keep them up to date.
4. *Integrating customer Web data with the third party aggregators.* To build better profiles of customers, it is often important to link the Web-based online data with the demographic data from the third party aggregators, such as Acxiom. Although important, such integration provides various challenges, including privacy challenges. Therefore, it is important to understand exactly which types of this third party aggregators' data can be used without violating consumer privacy and how this integration can be done in the safest possible manner.

## 5 CONCLUSION

Over the last decade there are numerous of achievements that the field of Web mining has accomplished, it is a good time to think about novel promising areas that will advance Web mining in the near future. In this paper, we believe that CRM is such an area that can benefit from and contribute to further advancement of Web mining research. This is the case because CRM is an under explored field that has many open and interesting problems that are important for the computational researchers. In this paper, we discussed how and why Web mining is an important component of CRM and how some of its techniques can help advancing the field of CRM further. At the same time, CRM problems can also impose certain additional requirements on Web mining which may lead to the development of novel Web mining methods.

## ACKNOWLEDGEMENTS

This research was partly supported by the Department of Education Research Projects of Hainan Province (Project no.: Hjkj2012-61) and National Natural Science Foundation of China (NSFC, Project no.: 61075063).

## REFERENCES

Adomavicius G & Tuzhilin A (2001) Using data mining methods to build customer profiles. *IEEE Comput 34(2), February*.

Etzioni O (1996) The World Wide Web: Quagmire or Gold Mine? Comm. *ACM 39(11), 65–68*.

Geoffrion A & Krishnan R (2001) Prospects of operations research in the e-business era. *INTERFACES Customer relationship management and Web mining: the next frontier 611*.

Jiang T & Tuzhilin A (2006) Segmenting customers from populations to individuals: does 1-to-1 keep your customers forever? *IEEE Trans Knowl Data Eng 18(10), October*.

Kotler P (2004) The view from here. *In: [PR04], pp 11–13*.

Ling C & Li C (2001) Data mining for direct marketing: problems and solutions. In: Proceeding of 4th international conference on knowledge discovery and data mining. AAAI Press, New York, pp 73–79.

Ngai E et al. (2009) Application of data mining techniques in customer relationship management: a literature review and classification. *Expert Syst Appl 36:2592–2602*.

Nasraoui O et al. (2008) A Web usage mining framework for mining evolving user profiles in dynamic Web sites. *IEEE Trans Knowl Data Eng 20(2), February*.

Peppers D & Rogers M (2004) Managing customer relationships: a strategic framework. *Wiley, Hoboken*.

Pang B & Lee L (2008) Opinion mining and sentiment analysis. *Found Tends Inform Retr 2(1), January*.

Padmanabhan B et al. (2001) Personalization from incomplete data: what you don't know can hurt you. *In: Proceedings of the ACM SIGKDD conference on knowledge discovery and data mining*.

Palmisano C et al. (2008) Using context to improve predictive modeling of customers in personalization applications. *IEEE Trans Knowl Data Eng 20(11), November*.

Swift R (2001) Accelerating customer relationships. *Prentice Hall PTR*.

Wedel M & Kamakura W (2000) Market segmentation: conceptual andmethodological foundations, *2nd edn. Kluwer Publishers, Boston*.

*Information Systems and Computing Technology – Zhang & Gu (eds)*


# Study on loan interest rates risk of commercial banks based on adverse selection

Jinhong Wang
*School of Management Tianjin University, Tianjin, China*

ABSTRACT: Since the implementation of interest rate liberalization, Chinese commercial banks loan rates are allowed to fluctuate around the benchmark rate based on the range 0.9–1.9 times. Except for housing loans rate slightly below the benchmark interest rate, others are higher than it, the highest for the interest rate broke surface 90%, which reaches loan rates upper floating ceiling of central bank. Smaller interest margin may sink enterprises into troubles, including high financing cost, low profitability and even loss, thus undermining their willingness and capacity for loan repayment and making loans more risky. Through the analysis of the relationship between loan interest rate and expected revenue of commercial banks in the model of adverse selection, this essay attempts to arrive at the conclusion that high loan rates unnecessarily yield fairer returns and may even incur the risk of the inability to bring back principals and interests, and encourage commercial banks to pay greater attention to the establishment of loan interest rates and reduce risks accordingly.

## 1 INTRODUCTION

The sixteen listed Chinese banks' net profit for the first quarter of 2013 increased by 12.6% year-on-year to CNY 309.1 billion with daily earnings of CNY 3.4 billion, delivering an outstanding performance well above expectation. However, accompanied with this achievement was the constant rise in Non-Performing Loan (NPL) volume. The Industrial and Commercial Bank of China, who has earned the most profit in the first quarter, maintained a NPL ratio of 0.87% which featured an increase of 0.02% as compared to last year. Though the relative NPL ratio was within a controllable range, the absolute amount it represented was rather alarming. The NPL volume kept growing because of combined factors, such as the slowdown of economic development, the decline of the enterprise profitability, the increasing financing costs caused by the up-going loan interest rates which resulted in lower profit of the enterprises and their unwillingness for loan repayment. Under current economic situation, the increasing loan interest rate has shown a great impact on the forming of NPL.

## 2 EFFECTS OF LOAN INTEREST RATE INCREASE

The expected revenue of commercial banks relies on loan scale and loan interest rate, while commercial banks always pay more attention to loan risks. Credit rationing refers to that, considering the adverse selection and moral hazard of borrowers, lenders will exclude some risky loan applicants. Commercial banks can not accurately get the borrowers investment risks because of information asymmetry, so they often solve the problem by raising loan interest rates, which, however, unintentionally denies low-risk borrowers. Other wise, commercial banks tend to entice borrowers to venture in risky projects for higher revenues to cover their loan and related interests. The above two points are so called adverse selection which may lead to borrowers' moral hazards and increase the average loan risk of commercial banks. While higher loan interest rate may deflate low-risk borrowers' desire for loan application, while maintain loan access of risky borrowers who can accept higher interest rate but are less likely to pay back their loans [Holmstrom, B. and P. Milgrom, 1987]. That's why the NPL of commercial bank increases. Therefore, to increase loan interest rate may more likely to lower the expected revenues rather than boost that of commercial banks, since it increases the loan risks of commercial banks.

## 3 RELATIONSHIP MODEL OF LOAN INTEREST RATE AND EXPECTED REVENUE

Assume that one borrower has several investment projects that will turn out either success or failure. While $R_i > 0$ stands for the expected revenue of a successful project, $R_j$ represents that of a failed project. It is also assumed that congeneric projects yield the same average expected revenue as $T$ which is known to commercial banks through their experience in loan transactions. The project success probability is $\rho(R_i)R_i = T$, means the expected revenue a successful project brings is in inverse proportion to its success probability. Commercial banks determine their loan interest rate r based on the relationship between the average expected revenue $T$ and the success probability $\rho(R_i)$.

Assuming the fund that each project requires as $V_i$, the own fund that an enterprise distributes to the project as $v_i$, and assume that the enterprise gets a loan from commercial bank to process the project, then the expected profit of a successful project is as below:

$$R_i - (V_i + v_i)(1 + r)$$

P.s.: Since own fund involves opportunity cost, its loan interest rate must be taken into consideration.

The expected revenue of a failed project equals to below:

$$R_j - (V_i + v_i)(1 + r)$$

An enterprise won't apply loan from commercial bank if it estimates that the expected profit of the project is less than zero. Thus, the expected revenue of a failed project is zero, in other word, $R_j - (V_i + v_i)(1 + r)$ tend to reach zero [Holmstrom, B. and P. Milgrom, 1991].

Considering above two points, the enterprise's expected profit would be as below:

$$y = p(R_i - (V_i + v_i)(1 + r)) + (1 - p)\,0 = p(R_i - (V_i + v_i)(1 + r)) \quad (1)$$

No enterprise will invest on the project whose expected profit is zero. In this case, the own fund can be invested in other projects.

The first derivative of $p$ in function (1) is: $y' = R_i - (V_i + v_i)(1 + r) = 0$

This means that there exists a critical point in function $R^* = (V_i + v_i)(1 + r)$. Only when $R_i \geq R^*$ will an enterprise apply for loans. There, too, exists a success rate $p^*$, and an enterprise will only apply for loans if $p \leq p^*$.

$$p^* = \frac{T}{R^*} = \frac{T}{(V_i + v_i)(1 + r)} \quad (2)$$

Assume that the density function of $p$ on the closed interval [0, 1] is $f(p)$ and its distribution function is $F(p)$. The success rate of loan application is equal to:

$$\overline{p}(r) = \frac{\int_0^{p^*} pf(p)dp}{\int_0^{p^*} f(p)dp} = \frac{\int_0^{p^*} pf(p)dp}{F(p^*)} \quad (3)$$

The first derivative of $r$ in function (3) is:

$$\frac{\partial \overline{p}}{\partial r}=\frac{\frac{\partial p^*}{\partial r}p^* f(p^*)F(p^*)-\frac{\partial F(p^*)}{\partial r}\int_0^{p^*} pf(p)dp}{F^2(p^*)}$$
$$=-\frac{f(p^*)T(p^*F(p^*)-\int_0^{p^*} pf(p)dp}{F^2(p^*)(V_i+v_i)(1+r)^2}<0 \tag{4}$$

Function (4) signifies that, the higher the loan interest rate is, the lower the average quality of applying loan and the bigger the default risk will concern. Under the practical system of limited liability companies, expected revenue $R_j$ cannot be less than zero before costs are deducted. Enterprises can enjoy the benefits brought by successful projects, i.e. sharing profit and repaying principal with interest timely, without worrying about the loss from failed projects, e.g. partially or even totally exempt from repaying the principal with interest after expected revenue $R_j$ was brought in. Given the expected revenue $T$ of a specific project, a higher loan interest rate will lessen the profit y engendered by the project, leaving only the projects with potentially higher expected revenues qualified for loan application. And the fixed expected revenue $T$ also indicates that expected revenue have direct ratio with risks, but is in inverse proportion to success probability. Consequently, to increase loan interest rates will aggravate default risk by enabling risky projects to supplant their less risky counterparts [Meyer, M., T. Olsen and G. Tprsvik,1996].

$$\overline{\pi}(r)=\frac{\int_0^{p^*}\frac{T(1+r)}{(V_i+v_i)}pf(p)dp}{\int_0^{p^*} f(p)dp}=\frac{\frac{T(1+r)}{(V_i+v_i)}\int_0^{p^*} pf(p)dp}{F(p^*)}=\frac{T(1+r)}{(V_i+v_i)}\overline{p}(r) \tag{5}$$

Expected revenues of commercial banks not only depend on loan interest rates, they are also determined by borrowers' loan repayment rates. If a project runs successfully, the borrower always repays the principal and interest on time; on the contrary, a faltering project often leads to loan repayment default. Therefore, a rise in loan interest rate exerts an impact on the expected revenue. The influence function of expected revenue may not be monotonic, and the average expected revenue unit loan $\overline{\pi}(r)$ can also be related to the success rate of the project.

$$\overline{\pi}(r)=\frac{\int_0^{p^*}\frac{T(1+r)}{(V_i+v_i)}pf(p)dp}{\int_0^{p^*} f(p)dp}=\frac{\frac{T(1+r)}{(V_i+v_i)}\int_0^{p^*} pf(p)dp}{F(p^*)}=\frac{T(1+r)}{(V_i+v_i)}\overline{p}(r) \tag{6}$$

Function (6) represents that the average expected revenue $\overline{\pi}(r)$ equals the default-free expected revenue $T(1+r)/(V_i+v_i)$ multiplies the average default-free rate $\overline{p}(r)$. The first derivative of $r$ in function (6) is:

$$\frac{\partial \overline{\pi}}{\partial r}=\overline{p}(r)+\frac{T(1+r)\partial \overline{p}}{(V_i+v_i)\partial r} \tag{7}$$

$\overline{p}(r)>0$, so per unit rise in loan interest rate will directly add $\overline{p}(r)$ to expected revenue. $T(1+r)\partial \overline{p}/(V_i+v_i)\partial r$ represents the indirect risk effects caused by the rise in loan interest rate, showing that per unit rise in loan interest rate will lead to $\partial \overline{p}/\partial r<0$ unit rise in borrowers' default rate and $T(1+r)\partial \overline{p}/(V_i+v_i)\partial r$ unit decrease in expected revenue.

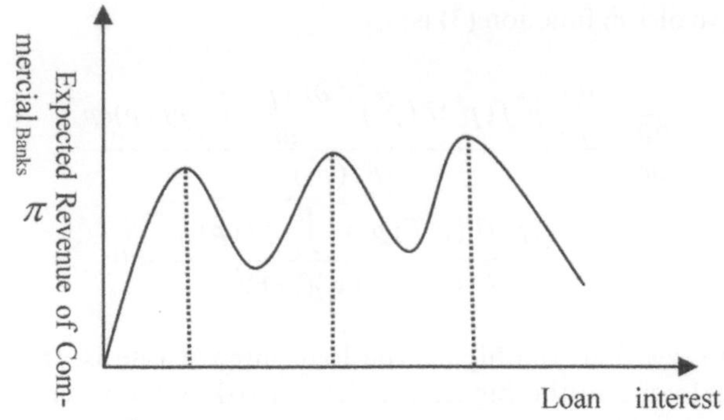


Picture I. The trend analysis graph of loan interest rate and expected revenue of commercial banks.

Therefore, commercial banks determine the interest rate level and risk of their loans. Due to information asymmetry, commercial banks can not accurately predict their loan risks and hence often raise their loan interest rates, which, however, unintentionally denies low-risk borrowers' access to loans. Adverse selection will result in credit rationing, thus enticing borrowers to venture in risky projects and increasing the moral hazard and average risk of commercial banks' loans. Borrowers who are willing to pay for high loan interest rates are the ones who are more prone to loan defaults. Therefore, the rise in loan interest rate may diminish commercial banks' expected revenues. As is illustrated in Picture I, it is more advisable for commercial banks to launch low-interest-rate loans, i.e. mortgage loans and hypothecated loan not fiduciary loan, and deny a proportion of loan applications, rather than approving most of them with their interest rates set high. [Raff, Daniel, 1995]

Obviously, the function of commercial banks' expected revenues may not be unimodal and allows multiple $r^*$ that can make $\partial\bar{\pi}/\partial r = 0$. For example, Picture I presents three such $r^*$. The exact number of $r^*$ depends on the strategy each venture participant adopts.

## 4 MODEL ANALYSIS

Assume that there are two types of projects, one is high risk type, and with a success rate $p = p_H$ and when the project is successful, the revenue is $R = R_H$, and the other type is low risk with a success rate $p = p_L$ and when the project is successful, the $R = R_L$. Obviously, the inequalities $p_H < p_L$ and $R_L < R_H$ are established. If loan interest rate $r \le r_a$, enterprises with both types of projects will apply for loans; if the loan interest rate $r \le r_b$, only enterprises with high risk projects will apply for loans; if loan interest rate $r > r_b$, no enterprise will apply for loans. It is also assumed that the proportion of high risk projects and less risky projects are $W_H$ and $W_L$ respectively, then $W_L = 1 - W_H$ and $K = T(1+r)/(V_i + v_i)$ hold. If commercial banks grant all loan applications, the function of expected revenues unit loan will be defined as:

$$\bar{\pi}(r) = \begin{cases} W_H p_H K(1+r) + (1-W_H) p_L K(1+r), \ldots\ldots\ r \le r_a \ldots ① \\ W_H p_H k(1+r), \ldots\ldots\ldots\ldots\ldots\ldots\ r_a < r \le r_b \ldots ② \\ 0, \ldots\ldots\ldots\ldots\ldots\ldots\ldots\ldots\ldots\ r > r_b \ldots ③ \end{cases}$$

The first derivative of function ① minus that of function ② comes to $p_L - W_H p_L = W_l p_L > 0$, signifying that the slope rate of function ① is higher than that of function ②. Picture II demonstrates the relationship between expected revenue and loan interest rate of commercial bank.

A. If loan interest rate $r \le r_a$, loan applications of all borrowers will be approved, and expected revenue of commercial bank is $\pi_2$, once loan interest rate $r = r_a$ undergoes a slight

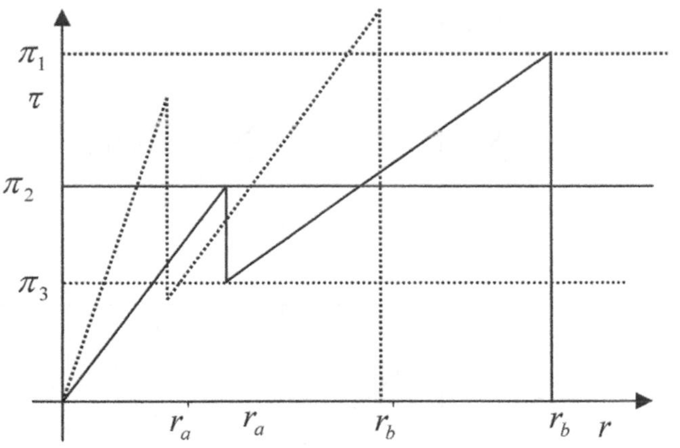


Picture II. Relationship between interest rate and expected revenue of commercial banks.

increase, low-risk projects will quit, thus causing commercial banks' expected revenue plunge to $\pi_3$. Some borrowers may consider about the continuity of business, so the project will be carried on even though loss has occurred, the average expected revenue of commercial banks will be $\pi_3 \leq \overline{\pi} \leq \pi_2$.

B. Assume loan interest rate $r_a < r \leq r_b$. Low-risk borrowers will withdraw their loan applications, while their risky counterparts will still be able to gain loans. Thus, the expected revenue of commercial banks will be $\pi_1 > \pi_2 > \pi_3$. Theoretically speaking, commercial banks will raise their loan interest rates to yield better expected revenues in this case. However, raised loan interest rates will also fan dangers that commercial banks are reluctant to take, such as increasing the average risk and default rate of loan. If loan interest rate $r < r_a$, commercial banks should raise their loan interest rates to extend their expected revenues. If loan interest rate $r > r_a$, while the expected revenue of a particular project will remain constant, the loan risk will grow. Under this condition, commercial banks usually require borrowers to take necessary steps, such as providing guarantee and mortgage, to reduce risks and secure loan safety. [Martimort, D.,1996]

C. If commercial banks set their loan interest rate below the equilibrium rate, namely $r < r_a$, the expected revenue of commercial banks will witness a growth. However, this unnecessarily means that, lower loan interest rates will give rise to higher expected revenues for commercial banks. This is because expected revenue, the theoretically attainable number, is somewhat different from actual revenue, the total principals and interests commercial banks really receive, due to interest revenue, which involves multiple factors like loan interest rate, loan volume and loan risk.

## 5 EQUILIBRIUM ANALYSIS OF CREDIT RATIONING

Credit rationing concerns the influence of four variables, loan demand $D$, loan interest rate $r$, loan supply $S$ and commercial banks' expected revenue $\pi$. Loan demand $D$ relies on loan interest rate $r$; loan supply $S$ depends on commercial banks' average expected revenue $\overline{\pi}$; loan demand $D$ influences loan supply $S$, commercial banks' average expected revenue $\overline{\pi}$ is related to loan interest rate $r$. The relationship between the four variables can be illustrated in the four quadrants of the following coordinate plane.

A. The third quadrant illustrates the relationship between loan supply $S$ and commercial banks' average expected revenue $\overline{\pi}$. In a capital market with perfect competition, commercial banks' average expected revenue $\overline{\pi}$ equals their loan cost. In this situation, commercial banks will be reluctant to supply loans due to no interest margin, and some of them will even face problems like high loan cost and negative interest rate owing to intensive competition.

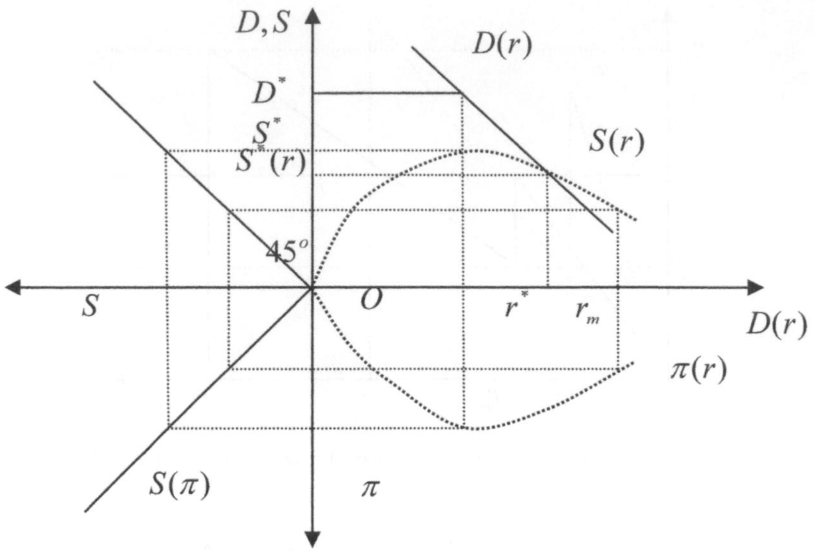


Picture III. Equilibrium schematic diagram of credit rationing.

Nevertheless, in response to government's macro control, commercial banks still will involuntarily supply this type of loan in less substantial volumes.

B. The first quadrant shows the curves of loan supply $S$ and loan demand $D$. If an increase occurs in loan interest rate, loan demand will dwindle despite commercial banks' willingness to extend their loan supply. Assume that $r^*$ represents the equilibrium loan interest rate featuring that loan supply and loan demand in the market meet and $r_m$ is the loan interest rate set by commercial banks. If $r_m < r^*$, loan applications of borrowers, risky and less risky alike, will be granted, and thus loan demand will experience an upward surge. In this situation, commercial banks will have to collect adequate capital to satisfy huge loan demand. However, they will also bear considerable risks, and their actual revenue will still be less than that of loan interest rate $r^*$. In contrast, if $r_m > r^*$, commercial banks will only satisfy the loan demand of risky borrowers and less risky borrowers who are willing to pay high interest rate. However, high interest rate will reduce commercial banks' expected revenue, and thus leading to the decrease in loan supply. Therefore, commercial banks will not supply such high-interest-rate loans with substantial risk, so their actual revenue will also be lower than that of loan interest rate $r^*$.

C. The fourth quadrant displays the relationship between commercial banks' average expected revenue unit loan $\overline{\pi}$ and loan interest rate $r$. If $r_m < r^*$, a rise in interest rate will grow commercial banks' expected revenue. On the contrary, if $r_m > r^*$, the rise in loan interest rate will lower commercial banks' expected revenue, which is in accordance with the analysis of the curves depicting loan supply, loan demand and loan interest rate.

## 6 CONCLUSION

If commercial banks' loan interest rate is lower than the market's equilibrium loan interest rate $r^*$, a rise in loan interest rate will suppress loan demand, raise commercial banks' expected and actual revenues and keep NPL risk within control. By contrast, if the loan interest rate of commercial banks is higher than the market's equilibrium interest rate $r^*$, a rise in loan interest rate will reduce loan demand and commercial banks' expected and actual revenues, and possibly make NPL risk out of control. Therefore, commercial banks should establish reasonable loan interest rates to reduce related risks, instead of taking raised loan interest rates that will increase both loan risks and their expected revenucs of the period.

## REFERENCES


Holmstrom, B. and P. Milgrom, 1987, "Aggregation and Linearity in the Provision of Intertemporal Incentives", Econometrical 55:303–328.

Holmstrom, B. and P. Milgrom, 1991, "Multi-task Principal-Agent Analyses: Incentive contracts, Asset Ownership and Job Design", Journal of Law, Economics and Organization 7:24–52.

Meyer, M., T. Olsen and G. Tprsvik, 1996, "Limited Intertemporal Commitment and Job Design", Journal of Economic Behavior and Organization, forthcoming. Discussion Paper No.102 (Sept. 1995), Nuffield College, Oxford.

Martimort, D., 1996, "Exclusive Dealing, Common Agency, and Multiprincipal Incentive Theory", Rand Journal of Economics 27:1–31.

Raff, Daniel, 1995, "The Puzzling Profusion of Compensation Systems in the Interwar Automobile Industry", in N. Lamoreaux and D. Raff (eds.) Coordination and Information: Historical Perspective on the Organization of Enterprise, Chicago University Press.

*Information Systems and Computing Technology – Zhang & Gu (eds)*
*© 2013 Taylor & Francis Group, London, ISBN 978-1-138-00115-2*

# A new network signal anti-collision algorithm

Xinchang Zhang & Moyi Duan
*School of Software Engineering, Nanjing Railway Vocational and Technical College, Nanjing, Jiangsu, China*

Yaping Ni
*China Medicine University, Nanjing, Jiangsu, China*

ABSTRACT: ZigBee network nodes in the great scale of data communication, communication does not agree for time domain by the conflict is put forward based on triangle calibration collision prediction of the channel conflict put signal mechanism. Use the same node of communication of the signal more than a signal virtual calibration area, through the establishment of the wireless sensor signal collision detection model conflict time domain prediction of communication conflict judgment. Through the time domain signals of 2 d regional collision forecast, calculation of the collision of the may probability. The experimental results show that the method can be used to big data, to node, long ZigBee communication network communication optimization, improve the efficiency of the node channel communications. At the same time, this method greatly reduces the wireless communication network data of the collision of the possibilities.

## 1 INTRODUCTION

ZigBee group of three types of networks: star, tree, mesh. Networking process is largely the same, but there is a big difference between the routing process.

ZigBee network is the first network initiated by the coordinator, scanning the environment whether there are other interference, choose a better channel and one PANID network.

Routing: routing node join the network, star network directly into the coordinator OVER; tree network routing node will find a better parent node, communication only after parent nodes, and so on; mesh network routing nodes randomly join the network, but the routing is in AODV mode, namely the on-demand routing, the source node information when initiating route discovery, the rest of the time just to periodic maintenance under the neighbor table.

ZigBee wireless sensor network is a new network based on new technology and computer technology, these years along with the related technology continues to mature, sensor technology, microelectronics technology, embedded technology and wireless communication technology. This calculation has also been widely used. With the rapid decrease the cost of wireless sensor nodes, wireless sensor network technology has been rapid development [Chen Xiang, Xue Xiaoping, Zhang Sidong. 2006]. The ZigBee wireless network technology because it can communicate between nodes, a self-organizing network system formed by means of wireless communication, the purpose is to cooperative sensing, collecting and processing in the coverage of network information. Acquisition and processing nodes can carry information, large area coverage, deployment flexibility is high, therefore become the most preferred. Because of its wide application and great business potential, therefore has become the focus of many scholars. But in some large-scale sensor network applications, with the increase of complexity, the number of nodes scale is also increasing, is prone to data communication conflict area, nodes if a conflict, to transmit data, resulting in energy waste at the same time, but also reduces the network communication efficiency.

In order to avoid these problems, this paper proposes a channel signal conflict mechanism based on triangle check collision prediction. Use the same node of communication signal more than a signal virtual calibration region, prediction of communication time conflict through wireless sensor signal collision detection model is established to judge. Predicted by 2-D local collision time, calculating the probability of collision may. The experimental results show that, using this method can carry on the optimization to the large data, long time of node, the ZigBee communication network, improve the efficiency of [Jiang Jianchun, Wan jade-like stone, Yi Gang, Hu Bin. 2011] node channel. At the same time, this method also greatly reduces the possibility of data collision of wireless communication networks.

## 2 DESIGN AND WORKING PRINCIPLE OF ZIGBEE NETWORK NODE

Communication in wireless sensor network is completed by nodes. A node can complete the sensor data acquisition, information processing, information remote transmission [Liu Dongsheng, Zou Xuecheng, Li Yongsheng et al. 2006].

The sensor node is complete the following functions are as follows.

### 2.1 *The information collection*

The effective information collection is the basis of the wireless sensor network, different information outside the formal by different types of sensors is completed, the first step in information processing. Due to the type of information collection is different, so the sensor types are varied. With the development of artificial intelligence and image processing technology, the application of the image sensor are more and more widely.

### 2.2 *The information processing*

Because the data format, attributes are not the same, so we must using information fusion technology is necessary, in order to ensure that the data collected in the minimum error, be handled effectively, ensure the accuracy of data transfer.

### 2.3 *The data transfer*

At the completion of the data processing, through the massive data transfer between nodes, node selection path the most convenient route, according to certain rules, so as to realize the accurate transmission of data.

In the effective transmission of data, along with the rapid increase of the number of the network nodes, the communication between nodes is not a single, simple form, although the nodes of the network is more and more complex, contact between the nodes of the communication and synchronization decreased. Communication between nodes independence more and more strong. At the same time, the communication between nodes of more and more. Therefore, node communication in conflict is inevitable, if at the same time, different nodes send data packets to the communication with a node at the same time, it is possible in the presence of conflict in the node. Once the conflict more data, then the nodes may be caused by energy depletion, delay in data transmission, communication between nodes under the efficiency problem, affecting the normal operation of.

With the rapid development of network, wireless sensor network has been widely used in many products. In the process of data transmission in eliminating conflicts although many, but still not well solve the existing node problem. In addition, although the node capacity and greater ease of sensor, data conflict problem to some extent, but with the transfer of data is more and more, the problem still is more and more outstanding.

Through the above elaboration content, need in local area network applications, compatible with other related data, and can reduce energy consumption, improve the speed of data processing, so as to realize the function of data transmission and reasonable.

## 3 TRIANGLE CHECK COLLISION PREDICTION OF ANTI-COLLISION ALGORITHM BASED ON ZIGBEE NETWORK

Because of the multiple nodes in the communication process, there may be a conflict in the time domain, on the same node sends data to different nodes, although different paths, but is likely to arrive at the same time. The data will have some conflict. Conflict exists effective detection, using the delay mechanism, routing protocol design, is an effective way to avoid conflict. Its principle is as follows.

### 3.1 *To establish virtual signal test range*

Using the Delaunay triangulation of most mainstream triangulation algorithm will detect node region scope division, this division can effectively prevent the node conflict, Delaunay triangulation algorithm is an efficient partitioning method can effectively prevent invalid triangle appears. To the Delaunay triangulation should satisfy some constraints for the following:

1. of three adjacent nodes in wireless sensor network as the basic point of each node is connected to form a triangle partition D region;
2. will not Delta region mapping sick in a central position;
3. change the node itself in the area will not affect the adjacent areas separated by region or point in the area.

If the constraints are partitioned areas in line with the above is the triangulation.

### 3.2 *Data collision discriminant*

Regional division of the ZigBee network should be carried out to prevent the conflict control, the traditional wireless sensor network to prevent communication conflict is through public channels are divided to different time gap to the control mechanism, when the communication node is too much will appear divided disturbance control failure disadvantage, this paper proposes a new control network the mechanism for conflict, the following principle.

After the first effective division of triangle area above the node will impact two-dimensional coordinates of the calculated. This paper assumes that the conflict exists and when the conflict occurs when the collision point must appear after triangulation later the area, so it is easy to calculate the two-dimensional coordinates of collision point. Collision node test scale can after several iterations to calculate the complete more than one region detection. If the triangle $\Delta A_k B_k C_k$ and $\Delta D_{k+1} E_k F_k$ is The wireless node region will be the test.

The above analysis, $\Delta A_k B_k C_k$ and $\Delta D_{k+1} E_k F_k$, Representing different collision detection area. Assuming the point $D_k$ Is the collision region $\Delta D_{k+1} E_k F_k$. In the two-dimensional coordinate mapping of the site. The site after the line after another is monitoring area $\Delta A_k B_k C_k$, So there will be such a relationship between the occurrence of regional location node regional conflict, can be $\underline{X}_F \cdot \underline{n}$ and $\underline{X}_D \cdot \underline{n}$. Symbols are compared, results show the conflict which a conflict class conflict, judgment belongs to the method which area to identify the specific as follows.

Assume the existence of vector equation

$$\begin{cases} \underline{X}_F = \underline{F}_k - \underline{A}_k \\ \underline{X}_D = \underline{D}_{k+1} - \underline{A}_k \end{cases} \quad \text{and} \quad \begin{cases} \underline{X}_F^n = (\underline{X}_F \cdot \underline{n})\underline{n} \\ \underline{X}_D^n = (\underline{X}_D \cdot \underline{n})\underline{n} \end{cases}$$

In the formula, the normal component $\underline{X}_F^{\prime n}, \underline{X}_D^{\prime n}$ and $\underline{X}_F, \underline{X}_D$ representing the network node signals in two-dimensional region has been mapped in the normal vector, represented by. To the node data of $\underline{X}_F \cdot \underline{n}$ and $\underline{X}_D \cdot \underline{n}$ to compare the same symbols, symbols that connect nodes data $D_k$ node data of $D_{k+1}$ may be through the region to be detected in $\Delta A_k B_k C_k$ ,indicating that the two regional data transmission conflicts may occur. Because the connection $D_k F_k$ is used as the node signal and another wireless network node must pass through $\Delta A_k B_k C_k$, it also has certain probability show traffic conflict. The same triangulation region after the edge may also be related to the Delta region other also exist certain collision probability. So even if the connection nodes $D_k$ and $D_{k+1}$ did not form and monitoring area of the intersection of $\Delta A_k B_k C_k$, $D_k F_k$ and $\Delta A_k B_k C_k$ formed the probability of conflict communication. But the probability of such conflicts compared with the former will be greatly reduced, saving energy in wireless sensor networks as a measure of network performance index is very important, in the calculation of the network's conflict, the conflict in the type of neglect.

### 3.3 *The error correction*

Calculation of certain error itself, hence the need for error proofing certain when calculating the final result, so to obtain more accurate results, the error is small. In the case of no errors, the results obtained through coordinate space projection transform to the two-dimensional condition still is a straight line in the mathematical model. Such a law can be calculated results error explanation and correction.

A plane of the linear equation was $x' = ky' + b$, $(x', y')$ represents the projection of a point without error in computer coordinates. Because the error of the distortion of the results to correction and compensation using the following formula:

$$x = x' + \Delta x$$
$$y = y' + \Delta y$$

$(x, y)$ said calculated without correction results in conflict situations, coordinate plane of said: so will the type into $x' = ky' + b$ available:

$$x - \Delta x = k(y - \Delta y) + b$$

Into the radial lens distortion correction model:

$$x - (x - x_0)(k_1 r^2 + k_2 r^4) = k(y - (y - y_0)(k_1 r^2 + k_2 r^4)) + b$$

In the formula, $(x_0, y_0)$ coordinates are calculated results in computer projection plane representation

$r = \sqrt{(x - x_0)^2 + (y - y_0)^2}$, $k_1, k_2$ represents the radial distortion coefficient, $k, b$ is assumed that the shape parameter line.

Under the arrangement of type available:

$$x = ky + \frac{(ky_0 - x_0)(k_1 r^2 + k_2 r^4) + b}{1 - k_1 r^2 - k_2 r^4} \quad \text{Let} \quad F = ky + \frac{(ky_0 - x_0)(k_1 r^2 + k_2 r^4) + b}{1 - k_1 r^2 - k_2 r^4}$$

Type $F$ equation is a nonlinear equation, the parameter equations of the partial derivative of $F$ due $k_1, k_2, x_0, y_0, k, b$, after transforming can be linearized equations. Linear transform as long as the guarantee that the results unchanged ultimately with error results will not change. This paper uses artificial increase the collision early results are infinite, so it can be the collision results locking, the center point of collision as the main value. This point is fixed, linear and parameter change. The F of a series of parameters of $k_1, k_2, k, b$ partial derivative:

$$a_1 = \frac{\partial F}{\partial k_1} = \frac{(ky_0 - x_0 + b)r^2}{(1 - k_1 r^2 - k_2 r^4)^2} \quad a_2 = \frac{\partial F}{\partial k_2} = \frac{(ky_0 - x_0 + b)r^4}{(1 - k_1 r^2 - k_2 r^4)^2}$$

$$a_3 = \frac{\partial F}{\partial k} = y + \frac{y_0(k_1 r^2 + k_2 r^4)}{1 - k_1 r^2 - k_2 r^4} \quad a_4 = \frac{\partial F}{\partial b} = \frac{1}{1 - k_1 r^2 - k_2 r^4}$$

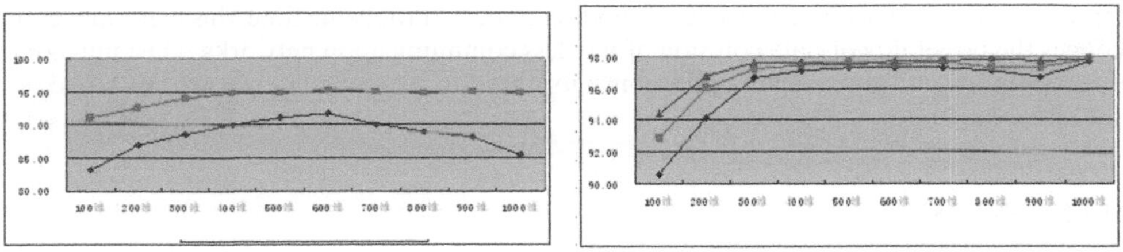

Figure 1. Node communication efficiency curve. (a) Contrast communication efficiency a node data (b) Contrast communication efficiency of large quantities of data.

The error equation is: $V = AX - L$

In the formula, $A = [a_1, a_2, a_3, a_4]$, $X = [\Delta k_1, \Delta k_2, \Delta k, \Delta b]^T$, $L = y - F^0$ the initialization of $k_1, k_2$ after the value is set to 0, $(x_0, y_0)$ as main point of initial results, $k, b$ can use mathematical calculations for fitting a straight line for $F^0$ is obtained through the initialization.

The above analysis shows that the error equation is a linear equation, the linear equation of the above can use the least squares method:

$$X = (A^T A)^{-1} A^T L$$

The corresponding linear if there is a $N$ point, coordinates of every point can be an error equation are derived, then the total there is a $N$ error equation, a collision calculation results of $M$ line error equation is: $V_{(M\times N)\times 1} = A_{(M\times N)\times 4} X_{4\times 1} - L_{(M\times N)\times 1}$, parameters is an iterative process. Through error compensation effectively, to optimize the results, to ensure the accuracy of the results.

## 4 EXPERIMENT RESULTS AND ANALYSIS

After data vibration sensor sampling, and communication. Wireless sensor nodes in a fixed can generate vibration ruler, slide ruler, the vibration, thus collected vibration data. The corresponding data connected through the serial port and PC simulation software. The SHM program for data communication and monitoring.

Using the background monitoring software, curve drawing transmission of a plurality of vibration data acquisition, in the experiment, data transfer volume compared with the traditional method and the method of the communication efficiency by using the transfer effect of measure the network data. Compared with the large amount of data and a small amount of data, the efficiency of network communication.

Seen from Figure 1 can be, when the vibration is increased, the amount of data increasing, nodes automatically shared transmission channel, will form the conflict, pictured above, above the curve for the network communication efficiency by this algorithm, efficiency is obtained by traditional method. After contrast can be seen, the communication mechanism that can well solve the conflict in communication, so the transmission efficiency will be higher than traditional methods.

## 5 CONCLUSION

This paper presents a channel signal conflict mechanism based on triangle check collision prediction. Use the same node of communication signal more than a signal virtual calibration region, prediction of communication time conflict through wireless sensor signal collision detection model is established to judge. Predicted by 2-D local collision time, calculating the probability of collision may. The experimental results show that, using this method can carry on the optimization to the data, the node, long time ZigBee network communication, improve the communication efficiency of node channel. At the same time, this method greatly reduces the possibility of data collision of wireless communication networks. The multi computer communication system anti-collision algorithm can be applied to many fields based on, and has great commercial value.

## REFERENCES


Chen Xiang, Xue Xiaoping, Zhang Sidong. The tag anti-collision algorithm of [J]. communication and information technology,. 2006 (220): 13–15.

Du Haitao, Xu Kunliang, Wang Weilian. The anti-collision algorithm returns the binary tree search [J]. Journal of Yunnan University (NATURAL SCIENCE EDITION) based on 2006.28 (01): 133–136.

Jiang Jianchun, Wan jade-like stone, Yi Gang, Hu Bin. And the realization of [J]. Computer measurement and control of 2011.7 communication system based on AUTOSAR framework.

Liu Dongsheng, Zou Xuecheng, Li Yongsheng et al. The collision algorithm [J]. Journal of Huazhong University of Science and Technology. 2006.34 in radio frequency identification system (9): 84–86.

Roberts, C.M. Radio Frequency Identification (RFID). Computers & Security, 2006.25 (1): 18–26.

Yang Ji, Chen Naigang. The multi computer communication of [J] and computer simulation of 2003.6 hardware platform of virtual reality, 77–79.

*Information Systems and Computing Technology – Zhang & Gu (eds)*
*© 2013 Taylor & Francis Group, London, ISBN 978-1-138-00115-2*

# The study of network node performance

Moyi Duan & Yingxi Wang
*School of Software Engineering, Nanjing Railway Vocational and Technical College, Nanjing, Jiangsu, China*

Yaping Ni
*China Medicine University, Nanjing, Jiangsu, China*

ABSTRACT: In order to mitigate the network performance by node failures, a novel availability evaluation algorithm (Traffic Queue Length based on Wavelet and Geom/Geom/1, TQLWG) is proposed based on previous studies. In this algorithm, the long-range dependence of traffic is decreased with wavelet transform at first, and the mathematic formulas of traffic queue length and node availability are established with Geom/Geom/1 model. Then, compared with the previous algorithm, a simulation was conducted to research on the relationship between the availability and influencing factors. The result shows that it is more adaptive for TQLWG.

## 1 INTRODUCTION

Along with the computer network scale increases, the effectiveness of the network nodes become the hotspot of current research. It points out the requirements of communication business performance to meet the failure conditions in parts of the lower network. At the same time related characteristics of actual flow also produced important influence for network system [Leland W. E., Taqqu M. S., Willinger W., et al., 1994]. In this regard, the domestic scholars have done a lot of work. In order to consider the communication network traffic factors and connectivity, presented in reference [Liu Aimin, Liu Youheng., 2002] reliability evaluation of communication network. Document [Sahinoglu M., Libby D. L., 2005] using probability expressions are evaluated the reliability of the network. Document [Lin Y. K., 2007] in network component failure of network performance by using probability weighted case, namely to meet the business requirements of the state probability summation. Document [Felemban E., Lee C. G., 2006] considers how to satisfy the requirements of reliability and distinction is not at the same time, ductility, especially on the network layer and the MAC layer at the same time to support differentiated services, the communication range, energy efficiency requirements. In addition, there are scholars reliability for two terminal and terminal put forward many model and the measurement of parameters of [Gebre B., Ram J., 2007].

At the same time, many researchers use to study the performance of network queuing theory. The literature of the [13] queueing system with negative customers M/G/1 vacation Bernoulli feedback. Steady-state queue length for Geo/G/1 document [Kim B., 2009] retrial queue system is studied, the characteristics of this system is that if the server is busy, customers into the retry stage, until the completion of services. The literature by considering the discrete time [Tang Yinghui, Huang Shujuan, Yun Xi., 2009] multiple vacation queueing system with batch arrival of Geomx/G/1, the study of transient characteristics and steady-state properties of captain. The literature [Tang Ying-hui, Yun Xi, Huang Shu-jun., 2008] the N—strategy of discrete time Geo/G/1 with the starting time of the queueing system, the transient distribution system captain discussed, the transient queue length distribution at any time recursive expressions are obtained.

Based on the above work, this paper presents a new characterization of nodes available algorithm TQLWG using the Geom/Geom/1 model (Traffic Queue Length based on Wavelet and Geom/Geom/1), the mathematical expression of the captain and the node flow validity is established, and the simulation results comparison and analysis of the factors affecting the flow performance.

## 2 THE NODE VALIDITY DEFINITION

The hypothesis network from node a to B multiple paths are up to, as shown in Figure 1. One, 1, 2, ..., m said the multiple paths and edge phase a, Max_out_flow (a) said node a maximum output flow, Input_flow (a) said node a the actual arrival flow captain, Out_flow (a, m) indicates that the actual output flow adjacent edges on captain M. Remove the A to the B office that a a phase limb still. The effectiveness of the node a had proposed the following definition:

*Definition* 1 in a network node a to remove the associated boundary m is effective for:

$$\delta(a,m) = \frac{\text{Max\_out\_flow}(a) - \text{Input\_flow}(a) - \text{Out\_flow}(a,m)}{\text{Out\_flow}(a,m)} \tag{1}$$

*Definition* 2 The effectiveness of a nodes can be defined as:

$$\delta(a) = E\left(\sum_{m} \delta(a,m)\right) \tag{2}$$

If the $0 \leq \delta(a) < 1$ side M can not or only partially be forwarded to other side, if $\delta(a) \geq 1$ m flow could all be the other side forwarding.

## 3 METHODS OF CALCULATING NODE AVAILABILITY

### 3.1 *The TQLWA algorithm*

Since the maximum output flow factors and node service rate, bandwidth, and the Max_out_flow in the network environment under the condition of fixed (a) values can be regarded as constant, so the key lies in the solution of Input_flow (a). But recent studies have found that the actual traffic exhibits long-range dependence obviously, so the combination of wavelet transform and ARMA model is proposed for the TQLWA (Traffic Queue Length based on Wavelet and ARMA) algorithm to study Input_flow (a), the specific algorithm is as follows:

1. When the initial time $t = 0$, the actual flow of node a Y wavelet Transform (T) decomposition, the wavelet coefficients and approximation coefficients of $A_j$ $D_j$;
2. Secondly, using wavelet coefficient $D_j$ ARMA model initialization, parameter estimation of $\alpha R(J)$ $(r = 1, 2, ..., P)$;
3. Is filtered by FIR filter, the output of approximate MA (q) process, and the estimation of parameter $\beta R(J)$ $(r = 0, 1, ..., Q)$;

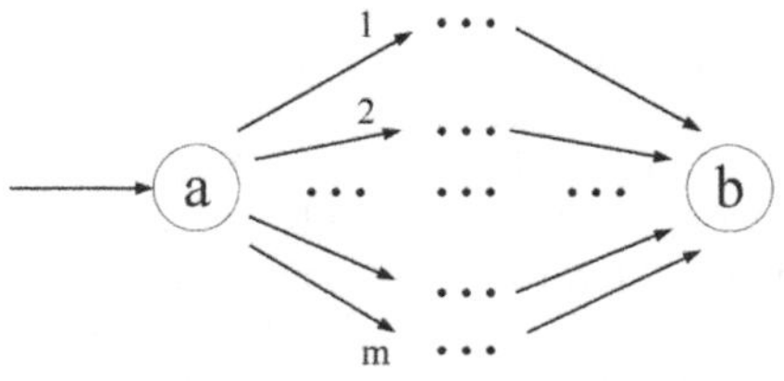


Figure 1. Simulation network topology.

4. Let $j = j+1$, skip to step (2), repeat the estimation of $\alpha R(J)$ and $\beta R(J)$, maximum level until the decomposition;
5. Based on the $\alpha R(J)$ and $\beta R(J)$ parameters of $P$ and $q$;
6. By the formula (3) calculating the wavelet coefficient $D'$ new $j$;

$$y(t) = -\sum_{r=1}^{p} \alpha_r y(t-r) + \sum_{r=0}^{q} \beta_r x(t-r), \quad \beta_0 = 1 \tag{3}$$

7. Finally, according to the wavelet coefficient $D'$ approximate coefficient $Aj$ and the calculation of the j, the reconstruction of the inverse wavelet transform, namely type (1) in the Input_flow (a);
8. By the formula (1) and (2) computing node availability of $\delta$a (a);
9. The algorithm terminates.

### 3.2 *The TQLWG algorithm*

But because of the limitation of ARMA model, this paper considers the flow of long-range dependence using wavelet transform, and then through the Geom/Geom/1 model to characterize the flow captain Input_flow (a) changes, this research is effective for node a. P assumptions that flow average arrival rate, service rate of a $\mu$ said node.

Let YT denote the business to t times the system flow, $\{yt, t \geq 0\}$ is a homogeneous MC, set the system transition probability $p_{ij}$:

$$p_{ij} = P\{y_t = j \mid y_{t-1} = i\} \tag{4}$$

Among them, $t \geq 1$. The system state conversion from I to j may have the following two situations: (*I*) ($T$, $t$+1) to reach a service flow, and ($t$–1, t) when $i$+1–$j$ flow away; (II) ($T$, $t$+1) when no flow arrival, and ($t$–1, $t$) with $I$–$J$ a flow away. Then the system transition probability of PIJ expressed as:

$$p_{ij} = pC_i^j \mu^{i-j} \bar{\mu}^j + pC_i^{j-1} \mu^{i+1-j} \bar{\mu}^{j-1} \tag{5}$$

The transition probability matrix can be expressed as:

$$\tilde{\mathrm{p}} = \begin{bmatrix} p_{00} & p_{01} & \cdots & \cdots & \cdots & \cdots \\ p_{10} & p_{11} & p_{12} & \cdots & \cdots & \cdots \\ \cdots & \cdots & \cdots & \cdots & \cdots & \cdots \\ p_{c-1,0} & p_{c-1,1} & p_{c-1,2} & \cdots & p_{c-1,c} & \cdots \\ \cdots & \cdots & \cdots & \cdots & \cdots & \cdots \\ \cdots & \cdots & \cdots & \cdots & \cdots & \cdots \end{bmatrix} \tag{6}$$

When $0 < p < 1$ and $0 < \mu < 1$, $\{yt, t \geq 0\}$ is irreducible, aperiodic [17]. Make $\prod = \{\pi_0, \pi_1\}$ Said distribution of stationary traffic captain, by the balance equation $\prod \tilde{\mathrm{P}} = \prod$ Can know:

$$\pi_k = \sum_{j=0}^{c+1} \pi_{j+k-1} p_{kj} \tag{7}$$

Among them, $k \geq 1$.
Make:

$$\{\pi_0, \pi_1\} = mb_c^{-1}\{b_0, b_1\} \tag{8}$$

The regularization term available:

$$m = \left( \frac{1}{b_c} \sum_{j=0}^{c-1} b_j + \frac{1}{1-\beta} \right)^{-1} = \left( \frac{1-b_c}{b_c} + \frac{1}{1-\beta} \right)^{-1} \tag{9}$$

Among them, $\beta$ (0, 1) is within a constant.
According to [17], the steady-state waiting time $W$ can be expressed as:

$$W = \sum_{k=0}^{\infty} p(W > k) = \frac{m}{(1-\beta)(1-(\mu\beta + \bar{\mu}))} \tag{10}$$

Node a steady state flow captain Input_flow (a) for:

$$Input\_flow(a) = E(y_t) = \frac{m\beta}{(1-\beta)^2} \tag{11}$$

According to the above method, the proposed TQLWG algorithm, is described as follows:

1. When the time $t = 0$, flow rate a initialization node arrival rate and service rate $\mu$ $p$;
2. Using the formula (12) shows the wavelet transform to the node a actual flow $y$ $(T)$ decomposition, wavelet coefficients of each layer are $D_{j,K}$ and approximate coefficients $A_{j,k}$;

$$\begin{cases} D_{j,k} = \dfrac{\sqrt{2}}{2}(A_{j+1,2k} - A_{j+1,2k+1}) \\ A_{j,k} = \dfrac{\sqrt{2}}{2}(A_{j+1,2k} + A_{j+1,2k+1}) \end{cases} \tag{12}$$

3. Using the reconstructed the inverse wavelet transform, to reduce the flow of Y (T) related to the long characteristics;
4. Based on the Geom/Geom/1 model of flow y $(T)$ to establish the transition probability matrix;
5. By the formula (9) and (11) to calculate flow captain Input_flow (a);
6. By the formula (1) and (2) computing node availability of $\delta$ a (a);
7. The algorithm terminates.

## 4 SIMULATION EXPERIMENTS

In order to validate the TQLWG algorithm, here set up network topology shown in Figure 1 in the NS2, long flow and the program generates the simulation verification. The simulation data based on the fractal Brown motion (Fractional Brownian Motion, FBM) model of [16], average arrival rate was 2279 kb/sec, the number of edges $m = 5$ node a association. At the same time, the simulation results in MATLAB analysis, evaluation of the effectiveness of the network node.

The first program produced 2000 long related data, the degree of correlation with Hurst (abbreviated as $H$) parameters that, when $H$ is (0.5, 1) means the flow with long-range dependence, where $H = 0.9$. The former 1000 data as prior information, we use wavelet transform (12) as shown in the recursive formula for decomposition of these 1000 data, to obtain the wavelet coefficients of $D_{j,K}$ and approximate coefficients $A_{j,K}$, where $k$ represents the number of wavelet coefficients in each layer.

The calculation of flow captain Input_flow based on TQLWA algorithm and TQLWG algorithm (a), and the combination of the 1000 data approximation coefficient prediction, based

Table 1. The algorithm error analysis.

| Algorithm | Error |
|---|---|
| TQLWA | 18.67 |
| TQLWG | 16.32 |

on the simulation results can be seen, long data TQLWG algorithm and program produces relatively close, indicating that the method has certain prediction accuracy. In Table 1 shows the error of the two algorithms.

In order to study the influence of the performance of the TQLWG algorithm, aiming at different $H$ values of queue length Input_flow (a) and the buffer is analyzed, from the overall trend of speaking, with the increase of Input_flow buffer (a) increased, and tends to be stable. Curve in a small buffer, smaller $H$ value corresponding to the Input_flow (a) larger; once the buffer exceeds 4000, the smaller $H$ value corresponding to the Input_flow (a) is small, curve mutation. The causes of this phenomenon lies in the traffic correlation on Input_flow (a) have a greater impact, the queueing performance of Geom/Geom/1 mutations based on. And can be reset effect by means of finite buffer and truncation effects explain. Replacement effect is that when the buffer is zero or empty, then the message the message and in front of the uncorrelated, thus weakening effect of long related, only when the message in the buffer is big, long-range effect, its impact is obvious. Similarly, the truncation effect of finite buffer refers to when the buffer is full, then the message is discarded, which will weaken the effect of long related.

## 5 CONCLUSION

The effectiveness of the network nodes, this paper on the basis of the previous work, we propose a new algorithm TQLWG using Geom/Geom/1 model. The algorithm firstly through wavelet transform to reduce the long correlation characteristics of traffic flow, the mathematical expression of the captain and the node of the derivation of Geom/Geom/1 model. And the simulation experiments to compare the performance of TQLWG algorithm and other algorithms, the results show that the algorithm has better adaptability. But the network availability is an important performance index, combined with network invulnerability and survivability modeling can be in the follow-up study, in order to form a more perfect evaluation system.

## REFERENCES

Albert R., Jeong H., Barabasi A. Error and attack tolerance of complex networks[J]. Nature, 2000, 406: 378–382.

Felemban E., Lee C.G., Ekici E. MMSPEED: multi-path multi-SPEED protocol for QoS guarantee of reliability and timeliness in wireless sensor networks[J]. IEEE Transactions Mobile Computing, 2006, 5(6): 738–754.

Gebre B., Ram J. Element substitution algorithm for general two-terminal network reliability analyses[J]. IIE Transactions, 2007, 39(3): 265–275.

Gunawan I. Redundant paths and reliability bounds in gamma networks[J]. Applied Mathematical Modeling, 2008, 32(4): 588–594.

Hardy G., Lucet C., Limnios N. K-terminal network reliability measures with binary decision diagrams[J]. IEEE Transactions on Reliability, 2007, 56(3): 506–515.

Kim B. Tail asymptotics for the queue size distribution in a discrete-time Geo/G/1 retrial queue[J]. Queueing System, 2009, 61: 243–254.

Kim Y.G., Shiravi A., Min P.S. Congestion prediction of self-similar network through parameter estimation network operations and management symposium[C]. Network Operations and Management Symposium, Piscataway, IEEE, 2006: 1–4.

Lazarou G.Y., Baca Julie, Frostv S., et al. Describing network traffic using the index of variability[J]. IEEE/ACM Transactions on Networking, 2009, 17(5): 1672–1683.

Leland W.E., Taqqu M.S., Willinger W., et al. On the self-similar nature of Ethernet traffic (extended version)[J]. IEEE/ACM Transactions on Networking, 1994, 2(1): 1–15.

Lin Y.K. Reliability of a computer network in case capacity weight varying with arcs, nodes and types of commodity[J]. Reliability Engineering and System Safety, 2007, 92(5): 646–652.

Liu Aimin, Liu Youheng. Analysis of communication network service performance of unreliable components [J]. Journal of electronics, 2002, 30(10): 1459–1462.

Paolo Crucitti, Vito Latora. Error and at tack tolerance of complex networks[J]. Physica A, 2004, 340: 388–394.

Sahinoglu M., Libby D.L. Measuring availability indexes with small samples for component and network reliability using the Sahinoglu-Libby probability model[J]. IEEE Transactions on Instrumentation and Measurement, 2005, 54(3): 1283–1295.

Satitsatian S., Kapur K. An algorithm for lower reliability bounds of multi-state two-terminal networks[J]. IEEE Transactions on Reliability, 2006, 55(2): 199–206.

Tang Yinghui, Huang Shujuan, Yun Xi. Captain [J]. Journal of electronic distribution of discrete time Geomx/G/1 queueing system with multiple vacations, 2009, 37(7): 1407–1411.

Tang Ying-hui, Yun Xi, Huang Shu-jun. Discrete-time Geox/G/1 queue with unreliable server and multiple adaptive delayed vacation[J]. Journal of Computational and Applied Mathematics, 2008, 220: 439–455.

WANG Jianwei, Rong Lili. Cascade-based attack vulnerability on the US power grid[J]. Safety Science, 2009, 47(10): 1332–1336.

Xiaolong Wu, Shahram Latifi. Substar reliability analysis in star networks[J]. Information Sciences, 2008, 178(10): 2337–2348.

*Information Systems and Computing Technology – Zhang & Gu (eds)*
*© 2013 Taylor & Francis Group, London, ISBN 978-1-138-00115-2*

# Author index

Printed and bound by CPI Group (UK) Ltd, Croydon, CR0 4YY

Printed and bound by CPI Group (UK) Ltd, Croydon, CR0 4YY
18/10/2024
01776251-0007